Ecology and Planning

An Introductory Study

Ecology and Planning

An Introductory Study

Paul H. Selman

GEORGE GODWIN LIMITED · LONDON

First published in Great Britain 1981 by
George Godwin Limited
the book publishing subsidiary of
The Builder Group
1-3 Pemberton Row, Fleet Street
London EC4P 4HL

British Library Cataloguing in Publication Data

Selman, Paul H.
    Ecology and planning.
    1. Land use — Planning
    2. Ecology
    1. Title
        333.7   HD 111

ISBN 0 7114 5555 4

Typeset by Alacrity Phototypesetters,
Banwell Castle, Weston-super-Mare and
printed and bound in Great Britain by
The Pitman Press, Bath

To Jill

Contents

Acknowledgements

Longman Group Ltd, 0.2; Earth Island Ltd 0.3; Wiley & Sons Ltd, 2.1; Office of the Governor, State of Oregon and Duke University Press, 2.7; The Planner 3.2 & 4; Scottish Geographical Magazine and Scottish Development Department 3.3; Journal of Environmental Management 5.1; Allen & Unwin Ltd, 5.4; Manchester University Press, 6.1; Macmillan 6.2, 3 & 4.

Tables: Leonard Hill 1.1; The Planner 1.2; Duckworth & Co. Ltd 5.1; Her Majesty's Stationery Office 5.2; RTPI Journal 5.3; David & Charles 5.5; Cumbria County Council 5.6; Journal of Environmental Management 5.7; The Planner 6.1.

Photos: The Countryside Commission 5, 10 & 18; Len Birks 7 & 16; The John Madin Design Group 15; Shell UK Ltd 17.

Figures

Tables

Plates

Foreword

This book has been written in the belief that ecology has a major contribution to make to planning, but that planners generally have little knowledge of the subject and see only a limited scope for its application. It is, therefore, introductory and synoptic in nature, seeking to demonstrate the relevance of ecological principles to a wide range of planning activities at a general, but hopefully not too superficial, level. I have not attempted to convert planners into 'instant' ecologists, but rather to suggest to them a variety of situations in which ecological advice may profitably be sought.

The major stumbling block to the integration of the two disciplines has been that of perceived relevance. Planners, and students of town planning, can readily grasp the pertinence of engineering, architecture, sociology and economics, but ecology, which entails the acquisition of unfamiliar scientific concepts and skills, often seems incomprehensible and appropriate only to esoteric, technical and rarely encountered problems. In order to overcome these difficulties, this book aims to establish a unifying ecological framework which embraces a whole gamut of 'environmental' issues, such as the urban fringe, energy conservation and development, pollution, rural land use, agricultural land loss and landscape science, and draws them into the mainstream of planning activity.

Inevitably, some of the terminology will be foreign to students of social science; rather than include a glossary, I have attempted to ensure that basic terms are defined as they are encountered in the text. There will still remain some unfamiliar terms, however, and I would recommend recourse to one of the excellent specialist dictionaries now available, such as: Michael Allaby, *Dictionary of the Environment*, 1979, Macmillan, and Alan Gilpin, *Dictionary of Environmental Terms*, 1976, Routledge and Kegan Paul.

Introduction

The place of ecology within planning

The attainment of a harmonious coexistence between the built and natural environments has always been a central pursuit of town planning. It was implicit in the pioneer housing and public health legislation of the 19th century, which prescribed that man should no longer be compelled to inhabit an environment full of squalor and disease. It was an argument re-echoed by Patrick Geddes, the progenitor of modern British town planning, whose education as a biologist led him to re-interpret the phenomena of urbanisation in ecological terms (1). Indeed, it is largely this concern for the general qualitative enhancement of the environment which has bestowed upon town planning its distinct professional identity.

The British, whilst they have long chosen to dwell in towns and cities, have, nonetheless, placed high value on the natural beauty of the countryside. It was no surprise, therefore, that they espoused with enthusiasm the 'environmental revolution' of the 1960s and 1970s; nor was it uncharacteristic that the Royal Town Planning Institute should lay claim to this movement and include within its syllabus a reference to 'planning conceived as applied ecology' (2)

Yet, in spite of this, ecology has generally been treated as a minor consideration in development issues — to be accommodated as and when economic circumstances permit. A consistent feature of most educational courses for planners is the manner in which 'ecology' tends to be rather summarily treated during the first or second year of a four-year course and thereafter not included in any systematic way. In local government matters are little different. An ecological input may be sought for a few isolated instances of site development or restoration, the importance of which in relation to physical, social or economic factors will probably be slight. But the possibility that ecology might have a more general relevance to the functioning of cities or to regional productivity is not seriously contemplated.

This book has been written in the belief that this broader view of ecology must be acknowledged by planners. Further, it contends that planning and associated environmental legislation provide an appropriate medium for the implementation of strategies for eco-development. It is directed at planning practitioners and under-graduates who require a general introduction to the subject and it neither purports to be a field manual for implementation nor a comprehensive treatment for every complex ecological problem. 'Planning' and 'ecology' are terms which are frequently misused. A few words should be said about their interpretation here. Whilst our major concern is with statutory town planning, a text which examines environmental management must inevitably acknowledge a broader definition of planning. Statutory planning itself embraces a number of nuances: in particular, it includes *programmes*, associated with the strategic, rational allocation of scarce resources, and physical plans concerned with *implementation*. Other theoreticians identify further spheres of planning activity, but this simple differentiation suits our present purposes well. Broad conservation objectives, such as reten-tion of the genetic diversity of our native wildlife stock or the quest for self-sufficiency, may become important inputs into strategic planning, whilst an appreciation of ecological principles is recognised as an essential ingredient for the success of environmental improvement schemes. In addition, we must broaden our traditional conception of the term 'planning', for any strategy of environmental management will require the cooperation of many other professions: leisure and amenity managers, environmental health officers, water authority officials and so forth.

The term 'ecology' has been broadened in two ways: to incorporate, within an ecological framework, selected topics from physical geo-graphy, a subject once extensively taught in planning schools but the status of which has now been seriously eroded by the introduction of other disciplines; and, more contentiously, to embrace a variety of 'environmental' issues which may collectively be thought of as 'human ecology'. The danger of thus broadening the scope of ecology is that its scientific rigour may become diluted; the danger of not doing so is that the ecological content of planning will remain peripheral and slight.

Human ecology – a legitimate concern of planning

In the course of this book it is intended to demonstrate that ecology, far from being a mere adjunct, should become a central theme of planning. Its status in planning considerations should be as elevated as, for instance, that of sociology or economics. Just as the relevance of those subjects arose from a critical consideration of past failures — namely that developments would founder if they ignored the aspirations and economic prosperity of the people for whom they were conceived — so ecology became topical following the widespread recognition of the limits to growth of human societies. A suitable starting point for our study is thus to consider the issues which gave rise to the so-called 'environmental revolution' (3).

Perhaps the most pervasive theme of modern environmentalism has been concern about the rate of human population growth. Demographic considerations have always been a feature of statutory plans, and most planning problems arise from the growth of human pressures on land, either as a result of growing numbers *per se*, or as a side-effect of the increasingly sophisticated demands of an existing population, or both.

This is a phenomenon with which some fairly clear biological analogies can be drawn. For example, cells grown in a culture increase in number by dividing every few minutes so that in a short time a few cells can grow to a population of millions. Thus, if a population were to start with one cell, its growth would proceed: 1,2,4,8,16,128,256, 512,1024 ... The initial slow increase in numbers is thus followed by a very rapid increase and if this is plotted on a graph the curve will rise almost vertically in its later stages. Since the trajectory of this curve is described by an equation containing a positive exponent, the pattern of growth is termed 'exponential' (Figure 0.1). Similar trends may be observed for the historical growth of human populations on a world scale (Table 0.1); at present the rate of increase for this is 1.9 per cent per year which means that population levels will tend to double approximately every 37 years.

Generally speaking, the intrinsic rate of natural increase for insects and wild animals is exponential. However, in nature, an upper limit exists to the size which a population can attain; as this limit is approached, so environmental resistance increases, and this gradually curbs the rate of growth so that numbers stabilise within that level which the environment is capable of sustaining (Figure 0.2). This limit, which is termed the 'carrying capacity', is determined largely by supplies of food, shelter and space (termed 'density dependent' factors) but also by factors unrelated to population size, such as climate and abundance of natural enemies (termed 'density independent' controls). The organisms inhabiting an environment thus do so in a state of competitive tension, which manifests itself as a dynamic equilibrium in delicate balance.

There has been a great deal of speculation as to whether or not the earth imposes an upper limit on the human population which it can support, and what the disastrous consequences might be of exceeding this carrying capacity. In 1798 the economist, Thomas Malthus, suggested that, since population grew exponentially and food production only linearly, our eventual fate would be starvation. He was not to

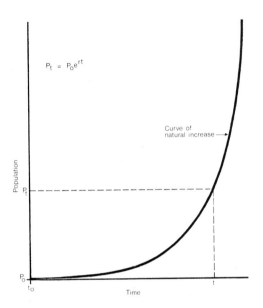

Figure 0.1   *'Exponential' population growth.*

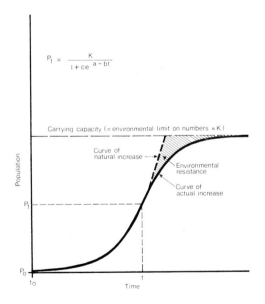

Figure 0.2   *'Sigmoid' growth, in which later rates of increase are curbed in response to environmental factors. (Adapted from N. Pears, 1977 Basic Biogeography, Longman.)*

Population size in millions	Approximate year	Years taken to increase population by 100 millions
1000	1830	900 000
2000	1930	100
3000	1960	30
4000	1975	15

Table 0.1   Natural increase of global population.

have foreseen that the capacity of the earth to sustain people could be raised dramatically by changes in technology. New strains of food, new manufacturing processes and the ability to exploit novel resources such as oil and uranium, for instance, have enabled human numbers to increase whilst still affording greatly improved standards of living — at least in the industrialised nations. However, innovations such as these presuppose a continued supply of resources to fuel industry and to allow cities to expand. If this stock of resources is allowed to be depleted in proportion to urban and industrial consumption then the Malthusian spectre may yet re-emerge.

A major challenge to our perception of environmental limits was issued by a group of ecologists in a manifesto called 'Blueprint for Survival'. In it they argued that:

'... the principal defect of the industrial way of life is that it is not sustainable... sooner or later it will end... in one of two ways: either against our will, in a succession of famines, epidemics, social crises and war; or because we want it to... in a succession of thoughtful, humane and measured changes...' (4).

Planners of all persuasions were thus called upon to utilise their foresight, powers and idealism to direct society towards a politically, economically, socially and ecologically stable state. The content and logic of this study received severe criticism, not only because of the unsubstantiated manner in which ecological concepts were applied to human populations, but also for the political naivety of its programme for change. Nevertheless, in the final analysis, its central contention was difficult to refute and its impact on contemporary thought was considerable.

More rigorous attempts to define the limits to growth imposed upon global populations have also been made using computer based systems simulations (5). The advantages of the systems approach to forecasting lie in its supposed comprehensiveness, since it attempts to incorporate information on every component of the system under study, and in its inclusion of 'feedback loops' which ensure that the effects of the behaviour of sub-systems on each other are fully considered. Such an approach would seem ideally suited to forecasting the fate of humanity, dependent as we are on the continued providence of the many interactive resource sub-systems which comprise our environment.

The simplest of these 'world models' rested upon an extrapolation of current trends of population growth, pollution levels and resource consumption, and assumed no major change in the forces which, historically, have governed the world system (Figure 0.3). The results suggested that food supplies, industrial output and population would grow in an exponential fashion until industrial growth was compelled to slow down by a rapidly diminishing stock of resources. Although population, being a lagged variable, continued to increase for some time beyond peak industrial output, it too was eventually curbed by deteriorating medical services and dwindling food supplies. Even the more sophisticated versions gave little cause for optimism: the most innovative scenario, for instance, incorporated an enhanced resource situation (including 75 per cent recycling), a massive reduction in pollution levels, a doubling of land yields, and 'effective' universal methods of birth control. Yet even here a constant population was sustainable only temporarily, being finally decimated by a combination of resource depletion, declining food production and the inevitable, if delayed, accumulation of pollutants.

Even the most sophisticated models will, of course, find their critics. In the above case, it was argued that the computer simulation was purely deterministic and took no account of human inventiveness; consequently, any cataclysm inherent in the input data was relentlessly fulfilled. Moreover, it was held, the data which were used, although far more exhaustive than any collated for any previous exercise, could in no way be considered sufficiently comprehensive and reliable. A third objection, particularly pertinent in the environmental context, was that the systems approach tends to average out variables through the whole system; thus, for instance, data on pollution, which is essentially a localised phenomenon, were somewhat meaninglessly related to world background levels (6).

Nevertheless, enough people were convinced that there was sufficient logic underlying these models to warrant a purposive, planned redirection of human endeavour. If this was to be effected, it would involve a far-reaching appraisal of alternative technologies and political systems. With no little sense of idealism, many planners became attracted by this path, and saw their professional task as being the replacement of what O'Riordan has termed 'technocentric' courses of action by 'ecocentric' ones. (7).

Figure 0.3   The outcome of two scenarios forecasting future world population levels in relation to resource availability.

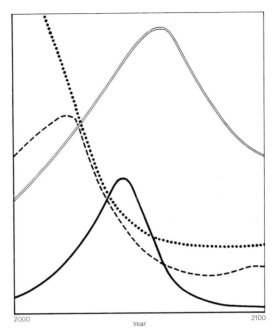

Figure 0.3a assumes no change in current patterns of consumption.

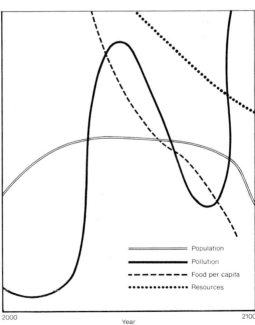

Figure 0.3b incorporates mineral recycling, pollution control, increased food production and effective birth control. (Adapted from Meadows/ Club of Rome (5).)

Towards a new environmental philosophy

The 'technocentric' mode of thought perceives the environment merely as neutral stuff from which man may profitably shape his destiny. This view has typified the conscious or subconscious aspirations of the majority in modern times. As a philosophy, it has remained seductive and plausible and, as evidence of its validity, one can point to a sustained increase in our material standards of living, the quality of our food supply and the time available for leisure.

Critics of the technocentric philosophy have pointed to the many spheres in which the social costs of production have now started to exceed the benefits of consumption and argue that if growth in demand is allowed to continue unchecked, an environmental catastrophe seems inevitable. In the industrial field, for example, supplies of essential energy resources and raw materials may become exhausted within a generation or two. Built-in obsolescence in consumer durables, and the widespread pollution caused during their manufacture, have in many places turned the environment into little more than a dump. In agriculture, enterprises have concentrated on such a restricted range of crops that monocultures have emerged, not only creating uniform habitats inimical to a variety of wildlife, but cultivating conditions in which undesirable species, previously held in check by their natural enemies, can multiply to pest proportions. Furthermore, the returns from intensive farming can only be sustained by substantial, costly inputs of energy, fertilisers and pesticides and, occasionally, by pursuing management practices detrimental to the long-term structure and fertility of the soil. These trends have been concealed from urban man, cocooned in his artificial habitat, but he also, often unwittingly, has eroded the rural resource through his contribution to urban sprawl and through recreational pressures which impinge upon ever more vulnerable reaches of the countryside.

Yet, by living off the *interest* of the earth rather than its *capital*, our requirements as a species could be met; our activities could complement rather than destroy the balance of nature. In an age when natural resources are becoming scarce and in which mankind increasingly displays a psychological need to establish his cultural roots and to achieve a sense of unity with nature, it could be contended that the 'ecocentric' approach — whereby we are seen as stewards of the environment conserving or enhancing its quality and productivity for the benefit of future generations — has become more relevant. Not surprisingly, it has been widely advocated that the planning profession should assume responsibility for effecting the change to ecocentric behaviour: one respected author has thus called for "... a purposive and regulated change to a design consciously formulated and reflecting man's highest aspirations" (8).

The most recent, and perhaps most significant, initiative to promote a 'regulated change' has come in the form of the *World Conservation Strategy* of the International Union for the Conservation of Nature and Natural Resources (IUCN). Lee Talbot, one of the leading architects of the Strategy, has stressed the need for a coordinated approach if its objectives are to be realised:

'...to date, most of our endeavours are fragmented, dealing with one or another problem in a largely isolated and consequently simplistic way, in the assumption that somehow one part of the environment was separate from, and could be dealt with apart from, the rest' (9).

The Strategy itself has triple objectives: to maintain essential ecological processes; to preserve genetic diversity; and to ensure the sustainable utilisation of species and ecosystems. It is essentially a global programme, oriented towards the major natural systems of the world, yet concerns itself also with the legislative and technical aspects of implementation. It is not anti-development, but argues that the development process has often proved *needlessly* destructive, generally as a result of inadequacies in environmental planning procedures. Although the town planning profession is not singled out as being of signal importance, it clearly has a key role to play, both in its own right, and as a contributor to multi-professional, interdisciplinary programmes for resource conservation and development.

Towards environmental action

Whereas the founders of modern town planning were visionary and ecocentric, contemporary practice often gives the impression of being procedural and technocentric. The ecological imagination has been ignored. There is a fragmentation of environmental action into 'urban' and 'rural', the latter perhaps being considered somewhat self-indulgent and less worthy of the committed planner's attention; it is with this sphere that ecology is generally, if sometimes erroneously, associated. The result has been compartmentalised management of the environment: town planners have been appointed to control building and engineering development, separate agencies have been established to ensure the development of farming, forestry and water supply, a conservancy has been formed to safeguard wildlife, harbour authorities supervise the coastal environment, various departments and inspectorates cover the living and working environments and national park authorities exercise a rather exclusivist watch over our grander scenic areas. Only belatedly have we, rather haltingly, moved towards corporate administration of resource protection and development.

It is hoped in this brief incursion into the fields of ecology and planning to reveal the ecological unity of the built and natural environments and to outline the scope which exists for tackling environmental problems in a comprehensive rather than isolated fashion. In order to provide a more even coverage of the total environment, less emphasis has been given to certain 'rural' topics, such as contemporary agriculture and forestry, which have pre-dominated in other accounts. Most importantly, it is the intention to re-assure those seeking to become involved in this field that there is, contrary to popular opinion, a range of techniques and legislative powers available for effective action.

NOTES

(1) Patrick Geddes (1886) *Conditions of the Progress of the Capitalist and of the Labourer*, Edinburgh, quoted in P. Boardman (1979) 'Challenge to human ecology', *Town and Country Planning*, 48 (6).

(2) Royal Town Planning Institute (1969), *Examination Handbook*, p. 25.

(3) M. Nicholson (1970) *The Environmental Revolution: a guide for the new masters of the world*, London, Hodder and Stoughton.

(4) E. Goldsmith, R. Allen, M. Allaby, J. Davoll and S. Lawrence (1972) *A Blueprint for Survival*. Ecologist, 2 (Also, Harmondsworth, Penguin).

(5) D. H. Meadows/Club of Rome (1972) *The Limits to Growth: a report for the Club of Rome's project on the predicament of mankind*, London, Earth Island Ltd.

(6) The major critique of the world models was: H. S. D. Cole, C. Freeman, M. Jahoda and K. L. R. Pavitt (1973) *Thinking about the Future: a critique of the Limits of Growth*, Sussex University Press.
A useful summary of the various arguments is contained in: Open University, Earth's Physical Resources Team (1974) *Implications: Limits to Growth?*, *Milton Keynes, Open University Press.*
Also, see K. Mellanby (1975) Can Britain Feed Itself?, London, Merlin.

(7) For a fuller discussion of 'technocentric' and 'ecocentric' philosophies see: T. O'Riordan (1976) *Environmentalism,* London, Pion.

(8) R. Arvill (1973) *Man and Environment: crisis and the strategy of choice*, 3rd ed. Harmondsworth, Penguin.

(9) A review of the World Conservation Strategy is contained in: R. Allen (1980), *How to Save the World*, London, Kogan Page.

1 The Abiotic System

Ecology is concerned with the study of the relationship between plants and animals and the environment in which they live. It is thus a unifying sub-discipline of biology, incorporating aspects not only of botany and zoology (the biotic system) but also of the complex of winds, currents, moisture, solar radiation and inorganic chemicals comprising the abiotic system which furnishes the economy of nature with its requirements of energy and materials. Incident solar radiation may be converted directly to food by photosynthesis: but it also provides the enormous quantities of energy required to drive the climatic and hydrological cycles. Similarly, the surface layers of the soil, constantly replenished by weathering and biological activity, represent a reservoir of minerals for plant life, whilst the earth's crust stores fossil fuels and ores available to subsidise human industry. It is essentially these subsidies which, on being released into the surface environment, yield surplus quantities of heat, energy and waste minerals and become a source of pollution.

Within the biotic system a distinct grouping of individuals of a given species, generally in sufficient numbers to sustain adequate genetic variety, may be referred to as a 'population'; a number of populations which interact, for instance via predator-prey relationships, or which in some other way share a relatively self-contained spatial unit, form a 'community'; the setting which a community inhabits, comprising as it does both biotic and abiotic elements, forms its 'environment'; and the combination of community and environment is called an 'ecosystem', which represents the principal synthetic element of ecological study and serves as a fundamental management unit in resource planning.

The Climatic System

The predictability of climate and regularity of the seasons form the fundamental controls on the growth and composition of ecosystems.

The mechanism of climate is one of perpetual motion, with the energy of the sun producing cyclic patterns in which the earth is alternately warmed by radiant solar energy and cooled as this is re-radiated into space, and in which precipitation falls to earth only to be returned to the atmosphere by evapo-transpiration.

Variations within elements of this dynamic system, especially those of air temperature, pressure and humidity, combine to produce a distinctive pattern of global climates. Thus, differential heating of the earth's surface produces characteristic zones of relatively high and relatively low pressure, with flows of air currents from the former to the latter giving rise to the main regional winds of the world. Superimposed upon this low-level circulation is a separate high-level movement of mainly westerly winds, which themselves underlie some semipermanent narrow sinuous currents of very fast moving air called 'jet streams'. Fluctuations in humidity may result in rainfall when the level of water vapour present exceeds that which may be contained at a given combination of air temperature and pressure (the 'saturation vapour pressure'). This condition may arise where air is cooled by being blown upwards over mountains, or where warm and humid converge with cold and rather dry air masses to form depressions. For droplets of rain to form, however, condensation nuclei must be present; these are commonly naturally occurring particles, such as pollen, but their numbers are greatly increased over urban areas where additional smoke and dust particles are emitted.

Within the first 80km or so of the atmosphere, a zonation may be distinguished according to vertical temperature differences. Thus, the lowest stratum is termed the troposphere, and is characterised by a fairly constant rate of decrease of about 2°C for every 300 metres (the 'normal lapse rate'), this cooling being achieved adiabatically, that is, by the expansion of an air mass rather than by its overall net loss of temperature. The troposphere is separated from the next layer, the stratosphere, by the tropopause, which is in turn overlain by the mesosphere and ionosphere, respectively. The vertical decrease in temperature which obtains under normal conditions produces a selfperpetuating circulation of the atmosphere, as warm air rises from ground level, cools on expansion in the less confined upper reaches of the troposphere and subsequently descends. However, the actual ambient change of air temperature with height, the environmental lapse rate, may depart from the dry adiabatic rate and temperatures may even, under certain circumstances, increase with height. This frequently occurs due to the rapid cooling of hard surfaces of built-up areas at night-time, or to the subsidence of cool air into valley bottoms, and is termed a 'temperature inversion'. Once this situation prevails, very stable conditions exist and the normal replenishment of fresh air at ground level fails to take place (1).

The seemingly inexorable changeability of this system, however, belies the immutable controls which govern its operation. First, although local heating of the earth takes place, there is no net warming up or cooling down of global mean temperatures, and this equilibrium is referred to as the 'heat balance'. Second, water vapour and ozone in the atmosphere restrict the harmful portions of incoming radiation from the sun, whilst clouds, in turn, tend to retard re-radiation of its heat, creating a 'greenhouse effect' which results in the maintenance of an equable climate at the earth's surface. Third, although its distribution is variable, the total amount of water present in the earth's surface, the atmosphere and the oceans is constant. And fourth, the constituents of the atmosphere — nitrogen (78 per cent by volume), oxygen (21 per cent), and traces of ozone, argon, carbon dioxide, and other, chemically inactive, gases — are, to all intents and purposes, fixed.

So far, man's activities have had relatively little effect on the macroclimate; even so, certain current trends could eventually prove significant. Concern has been expressed regarding increased concentrations of carbon dioxide in the atmosphere resulting from the burning of fossil fuels, the firing of vegetation and the operation of certain industries (2). Since carbon dioxide strongly absorbs outgoing terrestrial radiation in the infra-red wavebands, it can reasonably be assumed that global temperatures could rise, although it is uncertain whether this would be detrimental; it is even possible that there is a counter balance to this trend in the slightly cooling effect of higher levels of particulates in the atmosphere. Further concern has been aroused by the injection of water vapour into the atmosphere by high-flying aircraft, especially by supersonic vehicles, into the very dry stratosphere where even quite modest discharges could upset local water and heat balances. A similar source of disturbance is the emission by rockets and high-flying aircraft of oxides of nitrogen into the ozone layer — essential for the filtering out of dangerous short wavebands of solar radiation.

Changes in climate are not solely brought about by man's activities; on the basis of several sources of information it has been possible to construct a picture of naturally induced variations during recent geological history. In particular, the inference of the composition of former climax communities has been possible on the basis of pollen trapped in clay or peat deposits. The Quaternary Period was evidently characterised by a series of Ice Ages, the most recent having retreated approximately 15-20,000 years before present (ybp). Thereafter, the climate became warmer, during periods known as the Bölling and Alleröd interstadials, before cooling once more to produce the post-Alleröd recession. About 10,000 ybp our climate started to reach its warmest level since the Ice Age, this 'post-glacial climatic optimum'

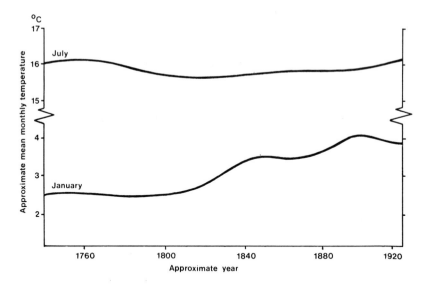

Figure 1.1 Generalised temperature trends in central England over a 200-year period. (Adapted from G. Sutton (undated) Weather and Climate, *BBC; Diagram by H. H. Lamb.)*

being divided into a dry period, the Boreal (9-7,000 ybp), and a wet period, the Atlantic (7-5,000 ybp); this was again followed by a further recession. Written records suggest that excellent climates probably prevailed during the Middle Ages, giving way to the 'little ice age' of 1550-1850, when the Thames frequently froze over (Figure 1.1).

There is now increasing concern that man's activities might be starting a climatic imbalance at more local levels (3). Perhaps the most evident are those microclimatic changes which have been in-duced in urban areas by artificial materials such as brick, concrete, glass and macadam. These effects are further compounded by the heat produced by metabolism and the various combustion processes, and by the changes in the chemical composition of the atmosphere arising from airborne emissions. Characteristically, daytime temperatures increase towards the centre of large built-up areas producing an 'urban heat island', although the reverse tends to be true at night. Several studies have shown that maximum daytime values tend to be about 8-9°C above surrounding ex-urban areas (4), with cellular patterns superimposed on this general trend, and thus whilst a 'climatological sheath' exists around single buildings producing variations in wind, temperature, humidity and soil moisture, a 'climatological dome' may envelop groups of structures.

The planner is pre-eminently concerned with the climate of open spaces within this environment, particularly the equable circulation of wind and the penetration of direct sunlight, both of which tend to be affected in a rather complex manner by the width-to-height ratios of buildings (5). In addition, the design and arrangement of buildings, by affecting local temperatures and winds, may exercise a considerable effect on air pollution concentrations (6). It is apparent that the inadvertent modifications of climates in urban areas has provided a number of advantages — such as fewer days with snow, a longer gardening season and lower heating bills — although these must be offset against the disadvantages of contaminated air, poorer visibility, higher rainfall, locally increased windspeeds and less sunshine. Indeed, so varied are the effects of human settlement upon microclimate that there is considerable weight in the arguments of those who favour the idea of air use planning on a metropolitan scale.

There has recently been speculation that we may presently be entering a renewed climatic deterioration, perhaps even a minor Ice Age, and there is a growing consensus of opinion that, at the very least, a phase of greater variability of both precipitation and temperature is under way (7). The principal effects of any such change are likely to be felt at the cold and dry margins of the cultivable areas of the earth, although a significant threat to middle latitudes does exist in the form of increased frequency of drought. More dramatically, continued warming of the polar regions could partially melt the ice caps, which contain enough water to raise sea levels by 60-80m. If we are indeed facing a future of more extreme and less predictable climatic conditions, it will create circumstances in which the ingenuity of planners in responding to a variety of unaccustomed contingencies will be seriously tested.

The Hydrologic System

Just as there exists a climatological cycle driven by the energy of the sun, so this is paralleled and partially overlapped by a hydrological cycle which transmits water through various pathways in the atmosphere, biosphere and lithosphere (Figure 1.2). Some 97 per cent of the water on the earth's surface is contained in the oceans; the remaining three per cent is fresh water, but the greater part of this is virtually immobilised in glaciers and ice caps, leaving less than one per cent in rivers, lakes and aquifers. Much of the water which falls as rain, sleet or snow will return directly to the sea. That which falls on land may be intercepted by vegetation or seep into the soil or rocks, whilst the remainder is collected on the earth's surface in expanses of standing (lentic) water, or enters courses of flowing (lotic) water, moving towards lower ground. This 'water budget' remains constant, only its quality changes: thus, as demands for water increase, supply programmes are tending to shift their emphasis from one of impoundment to one of protecting natural watercourses.

Plate 1.   The central area of Birmingham – a classic example of alteration of microclimate and hydrology by high density development. The River Tame to the north of the city frequently floods as water drains into it from the urban area at a rate faster than can be accommodated.

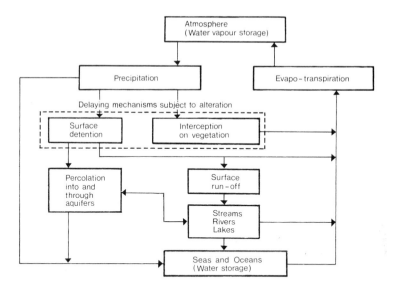

Figure 1.2   Principal elements of the hydrological cycle.

Several stages of this water cycle, since they may be markedly affected by alterations of the land surface, are worthy of particular attention by planners (8). Thus, whilst the natural controls of interception (by vegetation) and temporary storage of precipitation (by surface depressions) act as delaying mechanisms in the run-off process, the clearance of vegetation, or the replacement of an irregular by a smooth surface, will serve to reduce the time lag between heavy rainfall and flooding. The infiltration of rainfall into the ground serves both to replenish the water table and to control the proportions of rainfall contributing to surface or subsurface run-off, and the rate at which it occurs is determined by the nature of the surface cover and the permeability (and to a lesser extent the porosity) of the surface itself (Table 1.1). Surface run-off ensues when the intensity of rainfall exceeds the rate of infiltration. Its rate of travel and the consequent speed at which it causes a rise in the level of recipient watercourses is similarly influenced by the nature and amount of surface cover. Thus, where impermeability factors are increased there is a greater risk of peaky run-off and flooding.

Rate of flow in rivers is a complex function of slope, cross-sectional area, wetted perimeter and the roughness of the banks and bed; it may generally be noted, however, that obstruction of the flow will tend to produce greater depths of water for the same rate of flow and possibly lead to flooding. River channels are naturally adapted to accommodate the regular flood which occurs about once every one or two years (the 'bank-full discharge'), but any larger volume will cause inundation of the low-lying areas of the valley bottoms, these being termed the 'floodplains'. For economic reasons, a great deal of urban development has taken place on these sites, thereby increasing the potential damage likely to result from urban flooding, either by raising the level of the flood locally, or by aggravating natural flood problems downstream. Urban developments affecting stream flows tend to be of two sorts: those adjacent to the river, where impervious surfaces and their associated surface-water sewer systems lead to an increase in run-off and a decrease in the time-lag before its occurrence; and those which encroach into the channel itself, such as bridges, wharves and embankments, causing an impediment to river flows, and thereby raising the level of floods (9).

Type of Surface	Impermeability Factor
Watertight roof surfaces	0.7 to 0.95
Asphalt pavements	0.85 to 0.90
Consolidated heavy clay on restored land or construction sites	0.50
Macadam roadways	0.25 to 0.50
Gravel roadways and paths	0.15 to 0.30
Grassed areas and farmland, depending on slope and character of subsoil	0.05 to 0.25
Wooded areas depending on slope and character of subsoil	0.01 to 0.20

Table 1.1   Impermeability factors for principal types of ground cover in urban and rural areas. (Adapted from C. Tandy (8), based on data given by L. B. Escritt and Ministry of Agriculture.)

The most common response to the danger of flooding has been the construction of embankments, although these may produce unintended consequences, either increasing flood risks downstream or exacerbating the frequency of freak conditions in which floods exceed the height of embankments. In the long-term, only land use zoning can minimise the conflict between development and flooding, although it might lead to the sterilisation of tracts of riverine land. A more comprehensive approach to floodplain management and development control, which could coordinate the advantages of flat land, good soils and ease of communication afforded by floodplains with measures necessary to prevent excessive flood damage, has therefore been recommended (10). At present, applications for development in areas of flood risk are referred by the planning authority to the appropriate water authority; however, the process of referral is to some extent voluntary, and the planning department need not act on the advice given, an arrangement which has on occasion necessitated the expenditure of excessive sums of money on flood alleviation works.

It has been suggested that the water catchment in the uplands could form the overall context wherein the development of agriculture, forestry, recreation and, of course, water supply could be co-ordinated. The principal ecological factor in such an approach would lie in the maintenance of a suitable vegetation canopy; but, over the centuries, the uplands have been systematically denuded of their forests, whilst grass swards have been similarly changed in their species composition and continuity of cover. In many catchment areas, reafforestation has been carried out, both in order to reduce flood peaks and to ameliorate local microclimates by providing shelter from heat and wind. However, recent research appears to indicate that infiltration rates may be less satisfactory under coniferous forest than grassland and that, when water is scarce, such tree cover seems to produce higher rates of evapotranspiration.

A great deal of research remains to be done on the ways in which water cycle management and land use planning can interact. However, we may note that the development of urban areas tends to be associated with changes in water quality, with reduced infiltration rates to underground aquifers and with increased flooding risk, whilst land use changes in rural areas may markedly affect such elements as groundwater recharge, surface storage and soil erosion. The incorporation of hydrologic considerations in strategic forward land use planning and the fuller involvement of water authorities in the relevant aspects of development control are thus essential constituents of comprehensive environmental management.

The Soil System

Of all the criticisms which environmentalists have levelled against contemporary resource management practices, none has been more insistent than that regarding the possible destruction of the soil fabric by the uncontrolled use of inorganic fertilisers and heavy machinery. In addition, urban expansion has often depleted the extent and quality of our land resource by engulfing large tracts of soil high in agricultural potential.

In natural ecosystems, the destructive elements, creep and rainwash, are restricted by the climax vegetational cover, but agricultural land, with its highly modified standing crop, commonly has erosion rates five times the natural level. Wind erosion has similarly increased in recent years, particularly on the lighter soils, following the change from natural to inorganic fertilisers, the widespread practice of heavy mechanical cultivation and the removal of hedgerows, providing a greater 'fetch' for the wind (11).

Soil is often, rather pragmatically, defined as 'the stuff in which plants grow', a definition which highlights its properties as a natural medium for the rooting of and for the nutrient and water supply to vascular plants. It may, however, be rendered less suitable for plant growth where poor (usually compacted) structure hampers root development, where nutrients are scarce or are made unavailable by adverse conditions of acidity or alkalinity, or where poor permeability reduces the availability of water for plants.

By and large, soil is derived from the underlying parent material by a variety of evolutionary processes, although occasionally it can be transported from adjacent areas by the action of surface processes, in which case it is termed 'colluvial', or deposited from sluggish riverine or lacustrine waters in periods of flood, when it is termed 'alluvial'. Thus, soils are composed primarily of a skeleton of mineral material derived from the parent rock in the form of sand, silt or clay, and also usually contain organic debris which forms a surface humus layer and becomes intimately mixed in the upper 'horizons'. Some rudimentary soils, such as scree or blown sand, contain only mineral matter and are called 'skeletal' soils, whilst others, such as blanket peat, comprise almost solely organic material.

Perhaps the most important evolutionary process is that of weathering, in which essential nutrient 'cations' are released from the parent material by percolating rainwater, rendered slightly acidic by dissolved atmospheric carbon dioxide and by humic compounds derived from the surface litter. In cool, wet climates, the rate of weathering tends to slow down and so the leaching effect of rainwater, removing elements from the soil column, will tend to outstrip this, reducing the fertility of the soil. The carbonates, chlorides and sulphides, being highly soluble, are the first to be removed and thus many soils tend to be rather acidic. Under conditions favourable to intense leaching, and especially where the humus consists of pine needles or heather litter, 'podsolisation' may occur, in which even the far less readily soluble oxides of iron and aluminium ('sesquioxides') are mobilised from the upper horizons, leaving behind only a greyish layer of quartz granules, before being redeposited lower down the profile.

In broad terms, then, soil profiles tend to differentiate into distinguishable horizons, namely: a surface layer of litter, fermenting organic matter and humus; a layer from which minerals and bases are leached out (the 'A' horizon); a stratum into which some of these may be redeposited (the 'B' horizon); and the underlying parent material

Plate 2. Most soils are restricted in their use by a range of limiting factors. In this case, excessive stoniness is clearly a problem.

(the 'C' horizon). In Britain, the two principal zonal soils are the brown earth and podsol (alternative terms may be used in some modern classifications), the former usually formed on base rich rock and enjoying a southerly aspect at low altitudes, and the latter, more acidic and less satisfactory for agricultural purposes, but often suitable for forestry. Many intergrading soil types also occur, as do a number of local variants, such as the shallow 'rendzinas' on chalk parent materials.

Soil surveys and their accompanying memoirs, which describe the distribution and characteristics of a region's soils, generally record a uniform range of differentiating characteristics. Thus, the 'colour' of a soil may be used as an indication of its general state of aeration, with browns and reds showing oxidising conditions, whilst blues and greys suggest reducing, or 'glei', conditions. Second, the alkalinity or acidity of a soil — extreme values of which may render certain nutrients unavailable to plants — is determined by reference to its pH value (the negative exponent of hydrogen ion concentration) with values in excess of seven indicating alkalinity, and below seven, acidity. Third, 'texture' is used to refer to the relative proportions of different grades of material (sand, silt and clay), with a particular admixture of these, known as loam, generally being considered the most easily workable.

PART 1: Description of Soils

Series and symbol of selected soils	Texture and parent material	Soil group	Site	Land capability classification Class	Land capability classification Subclass	Limiting factors for soil use
Wetherby (Wu)	Loam or sandy loam	Rendzina	Hillocks and slopes	3	s	Shallowness and stoniness
Aberford (aF)	Loam or sandy loam	Brown calcareous soil	Undulating plateaux	2	s	Shallow in many places

PART 2: Physical attributes of soils which affect suitability for land use

Series and symbol of selected soils	Topsoil texture	Subsoil texture	Drainage class	Winter runoff potential	Permeability	Flooding risk	Bedrock Depth	Bedrock Nature	Slope (degrees)	Stoniness Topsoil	Stoniness Subsoil
Wetherby (Wu)	Sandy loam or loam	-	Free	Very low	Rapid	None	20-40 cm	Lower magnesian limestone	3-10	Stony	Solid limestone
Aberford (aF)	Sandy loam or loam	Sandy loam	Free	Very low	Rapid	None	40-80 cm	Lower magnesian limestone	1-8	Slightly stony	Slightly stony to stony

PART 3: Suitability of soils for selected uses; the two soil series are ranked relative to all the series surveyed

Selected uses Series and symbol of selected soils	Agriculture	Playing fields	Suitability as bulk filling	Building and road construction		
				Ease of shallow excavation	Suitability for cut slopes	Stability for foundations
Relevant factors in determining soil suitability (optimum characteristics in brackets)	Texture: (loam, sandy loam or silt loam) Drainage: (free to moderate) Permeability: (rapid but moderately retentive of moisture) Depth: (at least 80 cm)	Texture: (loam, sandy loam or silt loam) Drainage: (free) Permeability: (rapid) Slope: (very gentle, about 1 degree)	Texture: (well graded) Moisture content: (low) Plasticity: (low)	RD of soils: (low) Drainage: (free) Shear strength of rocks: (low) Undrained shear strength of soils: (low)	Drainage: free Permeability: (rapid) Drained and undrained shear strength: (high)	Drainage: free Permeability: (rapid) Drained and undrained shear strength: (high) Compressibility: (low)
Wetherby (Wu)	Rank: 7	Rank: 8	Rank: 1	Below water-table, Rank: 3 Above water-table, Rank: 4	Rank: 1	Rank: 1
Aberford (aF)	Rank: 4	Rank: 4	Rank: 1	Below water-table, Rank: 2 Above water-table, Rank: 3	Rank: 1	Rank: 1

Table 1.2 The assessment of soil capability for planning purposes. (Adapted from Casson et al (12).)

Fourth, soil 'structure' describes the solid part of the soil, ideally comprising rounded aggregates but often forming less satisfactory 'platy', 'blocky' or 'prismatic' elements which may make root penetration difficult, a condition sometimes aggravated by the passage of heavy machinery over poorly drained sites. Fifth, soil 'wetness', which is a function of a variety of climatic and topographical factors, will indicate where land drainage or, occasionally, supplementary irrigation is likely to be required. Finally, the 'cation exchange capacity', which is related to the chemically active fractions of the soil (clay and humus), refers to the ability of a soil to retain nutrient cations in such a manner that they will be available for uptake by plants, but cannot be leached out by percolating water.

It will be clear that soil surveys have traditionally emphasised agricultural criteria, and that they may consequently be of limited value for other land use planning purposes. Moreover, coverage of Britain is still very incomplete. Lately, there have been calls that, not only should the rate of survey be increased, but that the memoirs should record a wider range of information likely to be of use to other land use specialists, such as engineers, planners and landscape architects. For planning purposes, it has been suggested that a hierarchy of sources should be compiled, namely a soil map of the area concerned and single-feature maps based upon this showing the distribution of individual characteristics on which planning decisions have to be made; relative gradings of land potential could be derived from this for the conflicting uses under consideration. Where such data have been available, they have been found to assist greatly in planning decisions, particularly those relating to land development, waste disposal, recreational and amenity usage and highway construction (Table 1.2) (12).

The Geological System

The surface of the earth, from which we draw our mineral wealth, represents only a thin crust containing a finite amount of useful elements. This layer varies widely in its composition, but it is possible to make a broad distinction between the continental crust which is broadly granitic (sial) and the oceanic crust which is basaltic (sima) in nature. Beneath this outer shell lies the 'mantle', which is probably formed of peridotite, whilst at the interior lies a dense nickel-iron core. Under the intense temperatures and pressures which obtain in the mantle, the rocks tend to deform and 'flow' in types of convection currents. This in turn may trigger movements in the crust, causing earthquakes and other tectonic activity in marked bands across the earth's surface. These active bands transect the earth into a number of 'plates' which gradually move in relation to each other, perhaps at the

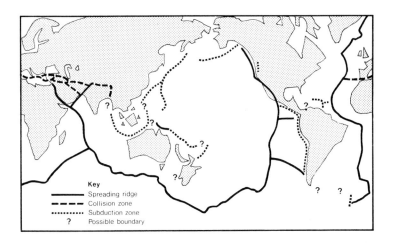

Figure 1.3 Major plate boundaries of the earth's crust – most of these are associated with active volcanoes or earthquakes. (Adapted from Geological Museum (1972) The Story of Earth, *HMSO.)*

rate of one or two centimetres per year, some spreading away from each other, whilst others collide with or are thrust below adjacent plates (Figure 1.3).

However, at the same time as these collisions and departures between plates cause the earth's surface to be forced up as fold mountains in some places, to be faulted laterally or vertically in others, or to sink as depressions elsewhere, the agents of erosion — ice, flowing water, wind and rain — act to level the ground once more, wearing down the higher land and transporting the sediment to basins where it will accumulate. Thus, rocks may be broadly divided into two groups — igneous or sedimentary — according to whether they were formed by tectonic processes and have crystallised from a molten magma, or whether they were deposited from a flowing medium and have lain *in situ* until consolidated. In addition, metamorphic rocks may occur, in which the original structure and composition of the rock have been altered by intense heat and pressure. Although sedimentary rocks would have been laid in horizontal strata, these may have been subsequently planed away by weathering, or tilted, folded or crushed by tectonic activity.

The processes which take place during the cooling of igneous rocks are particularly effective in concentrating certain types of rarer elements into mineral ores. As 'magmas' have cooled, often deep in the mantle, certain elements have proved unable to fit into the lattices of the common silicate-minerals because of differences in atomic structure, and these have become concentrated in residual fluids which crystallised out at a later stage. Alternatively, surface processes may lead to the concentration of ore deposits in sedimentary rocks, including the evaporation products of former seas. Similarly, coal strata represent the remains of tropical vegetation which decomposed in ancient swamps, initially producing peat, but which, with subsequent burial under other deposits, became more fully consolidated, with greater depth of burial increasing the 'rank' of the coal. Oil was formed from marine organisms in a comparable manner, except that it was produced offshore in a sedimentary basin from which, under pressure of burial, the deposits migrated along a stratum of permeable rock until being fortuitously trapped by some geological impediment, where the petroleum or gas formed a pool.

These processes are immeasurably slow, although certain tectonic activity, such as volcanoes or earthquakes, may be locally spectacular, and so the kinds of rocks present need bear no resemblance to modern climates or conditions: they may have been uplifted or downthrust far above or below their original place of formation, they may have been moved about the earth's surface by the drift of continental plates or by the action of transporting agents such as ice sheets, whilst rocks formed from magma deep beneath the crust may have been exposed at the exterior during aeons of 'planation'.

In Britain, a very wide range of geological environments is to be found (13). In much of Highland Scotland, North Wales, Cumbria and the South-West peninsula, igneous or metamorphic rocks, occur, often accompanied by workable ore deposits. Over much of the Midlands, Lancashire and the Trent Valley, the desert and lacustrine sands of the Permian and Triassic periods (270-170m ybp) have become consolidated into sandstone, as have those of the Devonian period (400-350m ybp) in parts of South-West England and South Wales. The Pennine and South Wales mountains are both associated with the blocky limestones, coal measures and resistant sandstones ('grit') of the Carboniferous (350-270m ybp), during which time a sea covered the area, bordered by tropical swamps. Parts of Eastern and North-Eastern England lay under the Zechstein Sea during the Permian which, on drying out, left behind evaporite deposits, notably potash in the Cleveland Hills. Marine conditions returned once more in the Jurassic and Cretaceous periods (170-70m ybp) depositing limestones, chemically precipitated iron ores, chalks and clays over the area east of an approximate crescent running from Teesside to Dorset. In the Tertiary period (70-12m ybp), sand, gravels and marine clays engulfed much of those deposits in the areas around the Solent and north-east, and west of London.

The remarkable degree of variation in Britain's scenery derives from this peculiarly diverse geological structure. Hard, resistant igneous and metamorphic rocks are associated with the mountain scenery on

Plate 3. Economic deposits of minerals are frequently associated with areas highly valued for their scenery and ecology. In this instance, flourspar is being mined at Youlgreave in the heart of the Peak District National Park.

the western and northern peripheries, whilst Carboniferous limestones and millstone grit form the Pennine backbone of England: it is in these areas that many of our National Parks and 'wilderness' areas are situated. The more recent limestone deposits, which display a characteristic 'scarp and vale' landscape, form many of our Areas of Outstanding Natural Beauty, whilst similar formations occur on the chalk downland of Southern England. However, it is an unfortunate fact of geology that often the most valuable ores and minerals are located within just such areas, for they tend to have been associated with the active phases of mountain building or with a rather restricted range of depositional environments. As a consequence, the exploitation of geological resources will, with predictable regularity, be surrounded by environmental controversy.

NOTES

(1) A straightforward account of temperature inversions may be found in: Open University, Environmental Control and Public Health Course Team (1975) *Air Pollution*, Milton Keynes, Open University Press, pp. 36-47.

(2) Reviewed by T. J. Chandler (1974) 'The management of climatic resources' in: A. Warren and F. B. Goldsmith (Eds.) *Conservation in Practice*, London, Wiley.

(3) For a general review of intentional and inadvertent modifications to the climate, refer to: Carrol L. Wilson (Ed.) (1971) *Report of the Study of Man's Impact on Climate*, (SMIC), Cambridge, Mass., MIT Press.

(4) A brief but comprehensive account of urban micro-climates is given by T. J. Chandler and S. Gregory (Eds.) (1976) *The Climate of the British Isles*, London, Longman.

(5) R. W. Gloyne (1974) 'Urban climates and micro-climates' in J. T. Coppock and C. B. Wilson (Eds.) *Environmental Quality*, Edinburgh, Scottish Academic Press.

(6) See, C. Wood (1976) *Town Planning and Pollution Control*, Manchester University Press, Chapter 5.

(7) The classic work has been carried out by H. H. Lamb: see, for instance, H. H. Lamb (1977) *Climate – present, past and future. Volume 2. Climatic History and the Future*, London, Methuen.
     Other useful reviews are: P. B. Wright (1976) 'Recent climatic change' in T. J. Chandler and S. Gregory (Eds.) op. cit. (4) and J. Gribbin (1978) *The Climatic Threat*, London, Fontana.

(8) An excellent account of applied hydrology is contained in C. Tandy (1975) *The Landscape of Industry*, London, Leonard Hill, Chapter 6.

(9) *Aspects of the theme of urban flooding have been the subject of several studies, for example: G. E. Hollis (1974) 'River management and urban flooding', in A. Warren and F. B. Goldsmith (Eds.), op. cit.* (2); A. J. Bowen (1972) 'The tidal regime of the River Thames: long term trends and their possible causes', *Philosophical Transactions of The Royal Society of London*, Series A. 272; M. D. Arnold (1976) 'Floods as man-made disasters', *Ecologist*, 6; and K. Smith and G. A. Tobin (1979) *Human Adjustment to Flood Hazards*, London, Longman.

(10) E. Penning-Rowsell and D. J. Parker (1974) 'Improving flood-plain development control', *Planner*, 60.

(11) An interesting summary of soil erosion and its control is contained in: A. Warren (1974) 'Managing the land', in A. Warren and F. B. Goldsmith (Eds.) op. cit. (2).

(12) J. Casson, R. Hartnup and R. Jarvis (1973) 'Soil surveys for integrated land use planning', *Planner*, 59. See also: K. W. Bauer (1973) 'The use of soils data in regional planning', *Geoderma*, 10.

(13) For a general introduction, see A. E. Trueman (1971) *Geology and Scenery in England and Wales*, (revised ed.) Harmondsworth, Penguin.

2 The Biotic System

In recent environmental studies man is increasingly included as one of the fauna, his lifestyle and needs reinterpreted in terms of the overall dynamic equilibrium of the biosphere. This broadening of the scope of ecology has proved controversial amongst purists who, with some justification, argue that man, who commands almost total dominance over nature, cannot be considered on a comparable basis with other species; others, conversely, maintain that it is this very capacity for environmental change which makes his inclusion essential. The planner can accommodate both these views, for, whilst matters of site management and planning rely on scientific ecology as a vital technical input, many strategic, sub-regional issues can be dealt with in broader ecological terms. Before ecology can contribute to land use planning in a fundamental manner, rather than merely as a technical adjunct, it is necessary for both planners and ecologists to accept the complementarity and equal validity of both viewpoints.

Ecology and Site Management

Succession

Probably the single most important ecological concept with respect to site management is that of vegetational succession. Casual observation will suggest that habitats relatively hostile to plant growth — such as skeletal soils or rock ledges — are colonised by only sparse and often structurally primitive flora. The 'pioneer' species, however, such as lichen and moss, assist soil development by contributing an organic fraction; long periods of further biological, physical and chemical weathering then create conditions more amenable to plant growth, so that a more advanced flora, in turn able to sustain a more varied fauna, may take root. Thus, early communities will be progressively invaded by more robust species which will successfully compete against those already there, probably leading to their extinction from the site. In this way the ecosystem gradually acquires a rich variety of species and develops a more complex organisational structure.

Assuming absence of climatic change or human disturbance, the final phase of this succession will be stable and will represent a characteristic vegetation type in keeping with the climatic and physiographic regimes of the region. This end-state is referred to as the 'climax' community, although modern ecologists generally refer to a 'polyclimax', since local variations may arise in response to differences in soil, drainage and so forth.

The application of the concept of succession to site management

The principles of succession suggest a number of practical applications. For instance, when a new land surface is artificially created, perhaps by ground moulding or by the deposition of industrial waste, a study of the natural processes of colonisation may reveal the best methods of establishing a plant cover on it. When the soil is sufficiently fertile to support a variety of species, a knowledge of the regional climax vegetation will indicate those native shrubs and trees likely to accelerate the establishment of a diverse, indigenous biota.

Such considerations are also critical in the use of land for recreation or nature studies. Very often, the earlier seral stages prove more attractive for such pursuits, and so some management input must be introduced to prevent the succession proceeding: lowland heaths, for instance, must be periodically burned or subjected to controlled grazing to prevent the invasion of birch scrub. However, these earlier stages — with their more open ground cover, less developed soils and more delicate flora — are especially sensitive to human pressure, and management must seek equally to restrict the adverse impacts of recreation (Figure 2.1) (1).

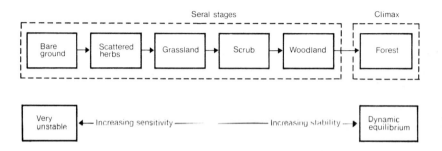

Figure 2.1   The vulnerability of an ecosystem to recreational pressure is largely related to its position in the ecological succession. (Source: Goldsmith (1).)

Erosion resulting from recreational usage may affect both the biotic and abiotic components of ecosystems. Most research on the former has concentrated on the effects of trampling, and confirms the increasing sensitivity of the earlier seral stages. A study of the Newmarket Heath racecourse for instance demonstrated a relationship between the varying levels of usage on different circuits and variations in the floristic composition in the grassland community (2). In general, such investigations indicate that vegetation tends to be bruised, that its height tends to be reduced, its flowering frequency decreased, whilst the more delicate herbs may be eliminated altogether. Additionally, soil compaction may restrict the abundance of soil microfauna, and animal life may be disturbed (3). The abiotic component of ecosystems may also be affected, especially where nutrient-poor or physically unfavourable soils are the limiting factors. A consistent field of attention has focussed upon sand dunes in this respect, especially the erosion of the 'yellow dunes' (those at the seaward edge of the dune

Plate 4.   The commencement of an ecological succession – fringing vegetation is becoming established at the margin of a recently formed lake.

system) by recreational pressure. If the pressure is intense, the fixed vegetaion cover, generally marram grass, may break up and the dunes start to blow inland. Re-vegetation is slow and even where artificially induced and adequately protected, might be expected to take several years. The most popular management technique has been to establish a pattern of zonal usage, whereby the public are channelled towards different sets of dunes on a rotational basis, providing alternate periods of pressure and recovery (4).

Recreational capacity

The concept of 'carrying capacity', the maximum population size which an environment is capable of sustaining in terms of its ability to provide essential resources, has also been applied to the maximum recreational pressure which a site can tolerate before it deteriorates. The practical application of this concept, however, is limited by the problems of interpretation which it poses. First, the term 'capacity' proves to be a multiple one: 'physical capacity' relates to the maximum number of users which can be physically accommodated on a site (eg cars on a car park); 'economic capacity' defines the threshold of usage which a commercial enterprise (such as a farm) can support before its profitability is reduced; 'perceptual capacity' refers to the level of crowding beyond which the recreational experience of individual visitors declines; whilst 'ecological capacity' alone identifies the maximum tolerable pressure before wildlife and vegetation are adversely affected (5). Second, it is difficult to define a single optimum state for an ecosystem: for some users even the slightest reduction in floristic diversity might seem inexcusable, whilst for others a carefully manicured, botanically uninteresting grass sward might be eminently acceptable. And finally, the response of vegetation to trampling is so lagged and so variable that it is virtually impossible to predict with accuracy and confidence, and thus the actual ecological capacity of any particular site must remain a matter of conjecture. This, however, does not mean that planners should dismiss ecological considerations as being imponderable, but rather that they should be more appreciative of the ecologist's difficulties in providing definitive answers.

Vegetation types

A further application of the understanding of succession and climax lies in the recognition of distinctive vegetation types, associated in turn with characteristic animal communities. These are commonly identified as woodland, farmland (grassland and arable), heath, wetlands, freshwater environments, coastal sites and artificial habitats

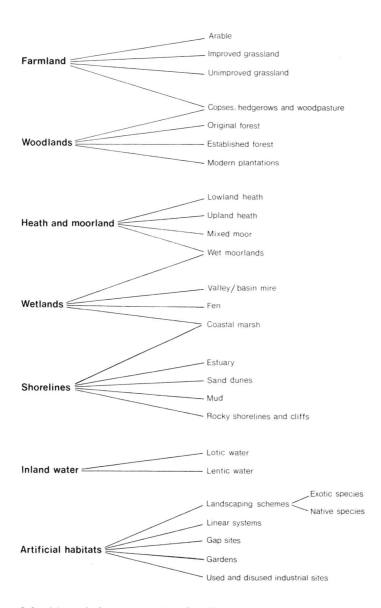

Figure 2.2   Major habitat types found in Britain.

(Figure 2.2). The study of the spatial variation of habitat types over the earth's surface is often referred to as 'biogeography'. A brief review of the major habitat types of temperate latitudes is of particular interest to us, for each will possess its own peculiarities of productivity and sensitivity, which are, for development and conservation planning the constraints upon which land use decisions must be based.

Plate 5. An illustration of woodland management by traditional methods – coppicing at Chalkney Wood, near Earls Colne, Essex.

Woodland

Woodland ecosystems possess unique characteristics by virtue of the sheer size and longevity of trees. Thus, the extent to which trees form a continuous canopy, intercepting sunlight and rainfall, has an overriding influence on the vegetational structure; their life cycle imposes a pattern of maturity, decay and natural regeneration, which ensures an environment of infinite variety and change. The 'stand cycle' (the period of growth from seedling to maturity) is a special feature of woodland: in a commercial plantation, where transplants are often set in a single operation, all the trees reach maturity simultaneously, whereas in a natural or semi-natural setting far greater diversity exists.

The characteristic structure of a woodland ecosystem comprises several layers, from the 'mainstorey' of tall trees, whose crowns form the general canopy, through the 'understorey' of medium-sized trees and shrubs to the herb layers and ground cover of mosses and liverworts. The presence of so many tiers, most notably in long-established deciduous woodlands, results in a great variety of ecological niches which provide ideal conditions for an abundance of wildlife. Coniferous forests typically possess fewer levels, due to the dominance of the mainstorey and the smothering blanket of acid needles which form the litter, and so will not sustain such a diversity of species. They may nonetheless be host to quite sizeable communities of certain animals and birds, especially if careful attention is paid to the detail of planting design.

Since prehistoric times British woodlands have been progressively cleared for agriculture, or plundered for their timber. The result is a prosperous lowland farming landscape in the south but in the north there is a legacy of bleak moorlands often little more than ecological deserts. Before the establishment of the Forestry Commission, only the medieval royal hunting forests and the modest estate plantations of the 18th and 19th centuries helped replenish the forest reserve to any great degree.

Of primary importance from a conservation viewpoint are the relics of native forest with a history of continuous tree cover since Atlantic times. Such sites are now extremely rare, the Caledonian pine forest, for instance, having been reduced to a mere 35 fragments, with a total area of only 10,000ha (6). Very often they form part of a commercially managed forest and, where this is the case, they may be suitable for designation as Forest Nature Reserves.

Also of considerable interest are secondary woods with native species. These often display a long history of management and although they may be floristically and faunistically poorer than original forest, they may possess a greater historical interest, reflecting in their species, composition and layout the rural economy of which they were a part (7).

Various traditional methods of management may be encountered in long-established woodlands, and (especially if the plantation is under the control of the local authority) it may frequently be feasible to continue these. Certain species may be 'coppiced', a practice in which the tree is felled and the stump allowed to send up shoots, providing an indefinite succession of crops of poles. Often, a proportion of 'standard' trees would be left to grow to maturity amongst these, ensuring a supply of timbers for larger constructional tasks. In areas where cattle were allowed to graze, trees were often 'pollarded', or lopped in the manner of a coppice above the height at which animals could reach the shoots.

The decline of our wood pasture, in which the art of growing trees amidst grazing livestock was practised, is of special concern to landscape planners in certain parts of the country (8). The skills of mixed woodland management had largely declined by the end of the 17th century and plantations after this date tend to comprise trees devoted solely to either amenity or commercial purposes. Planners have a particular interest in the preservation of the traditional English parkland landscape of the style associated with Kent and 'Capability' Brown which reached its zenith during the 18th century. In this the grounds surrounding the great country houses were landscaped so that pleasing vistas could be had from the house and gardens and also from vantage points around the demesne. Features in such parks often included vestiges of old hunting grounds, streams and lakes, irregular fields divided by hedgerows, and avenues and clusters of trees beneath which animals grazed. In recent years many such landscapes have lapsed into decline and various kinds of institutional or recreational usages have been suggested as means of saving their traditional character (9).

The scale of commercial planting has greatly increased during the 20th century, following the establishment of the Forestry Commission and the progressive mechanisation of foresty techniques. Thus, the new coniferous plantations of the uplands now form the most extensive type. Since these are areas of more limited conservation interest, they offer proportionately greater scope for recreational provision.

Creation of new woodlands is one area of planning in which an ecological input is most obviously essential. Where large-scale tree-planting is proposed, the observation of natural principles in respect of woodland composition and structure will greatly enhance the prospects of long-term success. An example of an ambitious planting scheme is provided by the landscaping initiated in conjunction with Oakwood residential development in Warrington new town. The area had previously been occupied by the Royal Ordnance Factory, and thus the soil was generally very poor and disturbed; however, an ecological approach, involving careful selection of tolerant native tree species, ensured that importation of large quantities of costly topsoil proved unnecessary. A web of woodland belts, laid down some three years in advance of the first housing completions, has been designed both to link with existing woods on the margin of the area and arranged so as to enclose cells for residential development and other land uses. The belts vary from ten to forty metres in width whilst, from their edges, narrow fingers extend into the built-up area, providing a continuous thread of nature through the urban environment and a linear habitat along which wildlife can spread. Principal (canopy) trees, understorey species, light-demanding species, and edge shrubs have been planted according to a variety of patterns, facilitating the development of a diverse woodland structure with robust and visually interesting edges. Although management costs have proved relatively low, careful timing and supervision of operations, based on detailed ecological knowledge, have proved necessary during the early stages.

Farmland

For centuries, wildlife has contrived to co-exist harmoniously with agriculture. The most notable contribution was made by the Enclosure Movement of the 18th century which resulted in a major increase in the number of hedgerows, although it was only the latest of a long series of trends which had left our countryside a varied matrix of meadows, boundary trees, small woodlands and common land. Since farming accounts for the use of some three-quarters of our land, farms, and in particular the trees, hedgerows, ponds and other semi-natural features which occur on them, must be counted as our most important nature reserve.

However, it has been claimed that modern agricultural methods are creating an environment incapable of supporting the former ecological diversity of farmland. Increasing mechanisation, reductions in farm manpower, amalgamation of holdings, changes in farm practice (notably the demise of rotational systems), changes in ownership and land values, and the removal of semi-natural cover from farmland — all apparently essential from the economic point of view — have produced an increasingly utilitarian and uniform farmscape. Heavy dressings of fertilisers, herbicides, fungicides and insecticides have likewise taken their toll, sometimes directly, by introducing toxins into food chains, sometimes indirectly, by unbalancing the productivity of ecosystems. Broadly speaking, three types of farming are to be found: lowland grasslands, hill and upland pasture and arable land.

In most areas beneath the tree line (ie the altitude above which tree growth is suppressed by severe environmental conditions) grassland is essentially a man-made vegetation type which has now become so dominant that it covers some three-fifths of our land surface. The animals which have grazed these swards have also markedly affected their ecology: during grazing, the sward is defoliated, enriched locally by faeces, and trampled, thereby inducing changes in its composition (10). Although a wide variety of grassland types may be identified in botanical terms, only three are recognised by agricultural statistics: permanent grasland (over seven years old), temporary grassland planted as part of an arable rotation, and rough grazing, which accounts for 60 per cent of agricultural grassland and includes hill pastures, lowland heaths, moors and unenclosed commons (11).

Much of the lowland grass is now an intensively grown crop and is of little ecological interest. The best agricultural return is achieved from a green uniformity — a reseeded, fertilised, herbicide-treated sward, dominated by a few particularly productive and nutritious grasses and clovers — whereas amenity and nature conservation value depend on the diversity of colour and form associated with the bygone pastures and meadows. However, fragments of a few important semi-natural communities of high conservation value remain. Perhaps chief among these are the botanically rich 'flood meadows'. Situated on broad floodplains, where seasonal inundation was often assisted by irrigation ditches, these were once actively manged for hay, but now, with the passing of horse transport, frequently lie neglected or have been reclaimed.

The botanical status of upland grasslands reflects on the one hand management practices and, on the other, a combination of elevational, climatic and soil factors. The upper, or montane, slopes afford generally meagre stocking rates, and grazing sheep selectively destroy the more interesting herbaceous plants. These areas support a poor and stunted acid grassland, although, occasionally, species-rich swards occur — as on the friable mica-schists of the Grampian Mountains.

The lower slopes present a greater variety of vegetation, the most characteristic form being the *Festuca-Agrostis* complex — consisting predominantly of sheep's fescue (*Fesuca ovina*) and the bents (*Agrostis spp.*) — and these pastures may be considered the mainstay of all grazing animals of British mountains. Unfortunately, overgrazing and poor management in many such areas have led to the more nutritious species being selectively removed, allowing unwelcome inundation by mat-grass, bracken and moor-rush (12).

Improvement of hill land has been in progress since the 19th century, although since 1947 its reclamation has been greatly facilitated by Government grants and, indeed, necessitated by the loss of adjacent land to forestry. Upland soils being generally acid, excessively wet and poor in essential nutrients require liming, drainage and heavy dressings of artificial fertilisers — operations which present problems on remote, steep slopes. In addition, improved land will normally be grazed on a rotational basis and so extensive and costly fencing must be erected. Commonly, amelioration involves the deep ploughing of heather moorland, and this has often been the subject of controversy on amenity grounds: however, equally striking landscape changes can be effected simply through chemical dressings, or by drilling seed together with a proprietary fertiliser mix directly into the ground ('sod seeding') (Figure 2.3) (13).

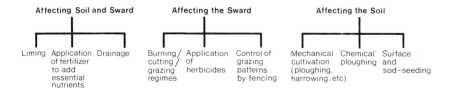

Figure 2.3 Principal methods of hill land improvement. (Adapted from McCreath (13).)

The report of the Public Inquiry into land management in Exmoor National Park noted that, if landscape controls were to be imposed by planners, the restriction of ploughing by itself would be inadequate and it would be necessary to specify stocking levels and maximum application rates of lime, fertilisers and chemicals (14).

The disappearance of many of the countryside's most diverse grasslands, and the increasing intensification of pastoral farming, have provoked widespread concern; but it is against arable farming that the most vigorous criticism has been levelled. In many parts of the lowlands, the emergent prairie-like landscapes, now increasingly devoid of their formerly extensive network of hedgerows, can support only a few 'steppe' species, such as the skylark and hare (15). Continuous cropping replaced traditional rotation systems of arable farming, in which livestock played an important role — now the animals are gone and with them the stout thorn hedge which formed their enclosure. The modern farmer not only finds little use for hedgerows but generally considers them as a nuisance; they restrict the efficient operation of large and costly machines, necessitating them to turn more frequently and to leave untouched wide swathes at the field margins. Where hedges are permitted to remain, mechanical trimming has often subjected them to a butchery from which many may never recover. The defenders of hedges, however, argue that in addition to their being valuable as wildlife refuges they can serve a useful purpose in the modern arable farm as shelter for crops and protection against soil erosion.

Plate 6. Modern agricultural practices have often rendered farmland visually and ecologically bleak.

A study sponsored by the Countryside Commission recommended that new cover should be established on sites awkward to plough, such as field corners, and that wherever possible cover should also be introduced to serve such practical purposes as shelter or game rearing. It was suggested that planting programmes could be facilitated by landscape agreements with landowners, fiscal incentives and the provision of financial aid to individuals wishing to plant trees for amenity purposes (16). These recommendations were not universally welcomed however and there were demands for stronger legislative measures to inhibit the removal of traditional farmland features; it was felt that the study had too readily accepted that such features were obsolete (17).

Whether the conservationist's concern is for the creation of new, or the retention of traditional landscape, it would be wrong to overlook the scope for compromise which exists even on the most efficient farms. Some measures which have been suggested include the alteration of the timing and intensity of chemical spraying, especially of wetlands, trimming hedgerows more carefully to allow some saplings to grow up, and following a pattern of hedgerow management.

Heathland

Heath is land dominated by the dwarf shrub, ling (*Calluna vulgaris*), although the bell and cross-leaved heathers may be locally more common. Wherever soils are acid, for instance in pine and birch woodland, *Calluna* has always been abundant, and thus forest clearance has allowed it to flourish and spread to open ground. Today, grazing and burning practices ensure that tree saplings are destroyed and that the heather cover is maintained (18).

Extensive heather moors remain in the east central Scottish Highlands, northern England and eastern Ireland; in the cool, moist climate of the west, ling thrives less well and it forms only one of several main constituents in the peat communities. The predominance of *Calluna*, and its influence on the microclimate and soils, limits the diversity of plant species, although some dwarf shrubs of the *Ericaceae* family are common, as are lichens and bryophytes, whilst a restricted but distinctive animal community is found, including several game birds and the mountain hare.

Heather moors support freely drained soils, and in many areas reclamation has taken place to produce highly productive farms and forests: even when unimproved, they are profitable for free-range grazing by domestic livestock. However, in northern Britain at least, their use for agriculture and silviculture is often subordinate to game sports, and where this is the case, grazing is generally insufficient to maintain a short, vigorous sward. Surplus dead and woody material must be removed by periodical burning in order to ensure a continuous supply of nutritious young shoots and to restore to the soil the nutrients locked up in the old plants. However, burning is a drastic treatment which, if not carefully controlled (and it often is not), can permit subsequent invasion by less desirable species.

In contrast to the uplands, only fragments of a once extensive lowland heath remain in the south of England. Once important to the agricultural economy and affording a variety of common rights, it has now been substantially transformed to permanent pasture, arable land, mineral working and forestry. Much of what is left is enclosed within military training areas, and whilst these severely restrict public access, they have at least provided some form of protection. The distribution of lowland heath has, however, become so fragmented that it now forms no more than a patchwork of isolated islands, and this has inevitably led to a reduction in the breeding success of many of its denizens, the fate of the near-extinct Dartford warbler having been particularly well documented (19). Where a management unit, such as a Country Park, has been established, the worst effects of fragmentation may be mitigated since a number of adjacent communities may be accommodated (20).

Wetlands

Today, as improved drainage systems become available and water-based recreation becomes more intensive, wetlands have become one of the most threatened habitat types (21). Often, wetlands depend for their conservation on the continuation of some former management practice: for instance, reedswamps, originally cultivated to provide thatching material, require the maintenance of a system of irrigation and drainage dykes, with periodical burning in order to increase bud density and to achieve evenness of growth.

Before the 17th century, fenland used to occupy much of eastern England, but since then it has been subject to systematic drainage so that true fen conditions are preserved — by very artificial means — only in a few nature reserves. The high lime content of fens contrasts markedly with most peats and thus they support an entirely different flora and fauna including many rarities. Bogs may develop where waterlogging restricts the full decomposition of organic debris, the result being a raw, acid organic soil, usually covered with mosses (especially *Sphagnum* spp.). The most distinctive types of community are 'valley bogs', found in shallow depressions among wet heathland, 'raised bogs', which develop on top of valley bogs by the continued accumulation of mosses, often producing a floating mat, and the 'blanket bog' of high rainfall areas (eg Rannoch Moor), where a raised bog engulfs the vegetation over wide areas.

Fenland, marshes and fringing vegetation represent early stages in vegetational successions and as such are likely to suffer recreational pressures. Therefore the planner, when considering development proposals, should seek detailed advice. The hazards are many: wet soils may be easily eroded, soil structure can be destroyed by trampling, boating can damage reedswamps and birdlife can be disturbed.

A less immediately obvious but equally serious loss of landscape quality has been caused by the extensive drainage of low-lying land in farming areas. Land prices having spiralled in recent years, farmers wanting to expand their holdings have resorted to the cheaper means of reclaiming tracts of wet ground. This has, of course, upset the ecological balance: riparian vegetation is removed, niches obliterated, whilst the lowering of the water table not only reduces the standing water and marshy areas available to wildlife but water chemistry may be affected. Underground drainage systems may further accelerate the movement of water containing fertilisers and pesticides into streams and lakes, and ditching operations tend to increase the sediment load of streams (22).

Despite the loss of many wetland habitats new sites continue to be established. Disused gravel pits (23), other abandoned industrial sites,

Plate 7. A lowland river with fringing wetlands presents a wide variety of niches to wildlife. Too often, habitats such as this are destroyed by channel-straightening operations.

reservoirs, although less likely to be of major conservation significance, can be made to serve the planner's purpose in deflecting recreational pressures away from more valuable areas.

Freshwater environments

Lentic, or standing, water occurs on the earth's surface as lakes or ponds (24). Lakes are commonly classified on a physical basis, according to their mode of formation, and may thus be circular (of volcanic or meteoric origin), sub-circular (of glacial origin), approximately rectangular (of tectonic origin), or crescent-shaped (ox-bow lakes formed by stream action). They may also be categorised ecologically according to nutrient content, with waters of low, high and medium nutrient status being termed *oligotrophic, eutrophic* and *mesotrophic*, respectively. Thus, alkaline or calcareous waters tend to be the most productive (for instance, where the watershed comprises chalk, limestone, schists, basic dykes, dolerites or gypsum), whereas those of acid response, which may have flowed over gneiss or granite, are generally low in nutrients.

As a lake ages, sediments and organic material gradually fill its basin and oligotrophic lakes may gradually evolve into shallow, gently-shelving eutrophic waters. However, the rate at which a lake becomes eutrophic may be accelerated artificially where organic wastes and nutrient-rich drainage from arable land discharge into it; this frequently leads to dense blooms of algae, oxygen depletion, and the elimination of freshwater species adapted to living in a medium with a low level of nutrient salts.

Lakes also possess varying characteristics according to their depth, which acts as a control upon the extent to which sunlight can penetrate. Thus, whilst all lakes have a shallow inshore 'littoral' zone, in which rooted plants can become established, deeper lakes may also display a 'limnetic' zone, which although lacking rooted plants, may support floating ones, and a deep-water or 'profundal' zone, in which photosynthesis is impossible. Lakes sufficiently deep to possess this last zone will probably develop vertical temperature differences during the summer as the surface layers become heated by the sun. This phenomenon is known as 'thermal stratification' and is typified by a warm upper layer (the *epilimnion*) overlying a cold lower one (the *hypolimnion*), the two being separated by a zone of rapid temperature decrease (the *thermocline*).

In contrast to the largely closed or self-contained systems of ponds and lakes, rivers and streams are open systems of lotic, or running, water. Whereas in a lake nutrient materials may be cycled several times, the inhabitants of a river must use material which remains at one point only very fleetingly. The biota of a stream thus display a number of adaptations for survival in varying conditions of turbulence and types of sediment. For example, in the fast-flowing upper reaches of a river are found animals able to wedge themselves into crevices or cling by means of hooks and suckers, together with small filamentous plants attached firmly to rocks; in the sluggish lower reaches fish with protrusible mouths able to suck up bottom deposits are accompanied by a greater abundance of rooted plants.

Canals display very similar conditions to sluggish streams, often possessing a wide variety of 'marginal' habitats. Little wildlife interest will be evident if the navigation is heavily used for commercial traffic. But under fairly light recreational and amenity usage the rate of sedimentation, and consequently of plant invasion, increases. Silting changes the shape of the channel and creates shallow shelves ideal for certain types of aquatic and semi-aquatic plants, particularly on the insides of bends where the scouring action of the flowing water is least. Although local pollution may be fairly severe, the oxygen content is generally quite high since aeration is helped by broken or closed lock gates which serve as weirs.

Shorelines

The shoreline is effectively a 'marginal' environment and, as such, exemplifies many of the conflicts characteristic of the interface between two contrasting habitats. It is a key area for intensive recreation pressure and, in addition, it attracts much industrial development, for the sea conveniently provides a source of cooling water, a receptor for effluents and a means of access for the import and export of materials.

One of the main habitats for environmental conflict has been the river estuary, and since over 30 per cent of the UK population now lives adjacent to estuaries, it is therefore not surprising that many are classed as 'grossly polluted' (25). The major pollutants are suspended solids, oil, hot water, toxic chemicals (for instance the cyanides and phenols which rendered extensive stretches of the Tees toxic to fish as early as 1931), heavy metals and substances causing de-oxygenation, such as untreated sewage and compounds discharged in a reduced state. In addition, the action of dredging may disturb the biota by increasing the channel volume, disturbing bottom-living species and affecting the mixing processes, sediment and turbidity.

The mingling of fresh and salt water in an estuary produces a sharp salinity gradient, varying from 0·5°/oo to 35°/oo dissolved salts, and this induces certain reactions which tend to flocculate mud and cause it to be deposited over large areas; for similar reasons, an accumulation of fine organic particles, known as 'detritus', also occurs, providing a substantial food source. Thus, whilst only a limited variety of species

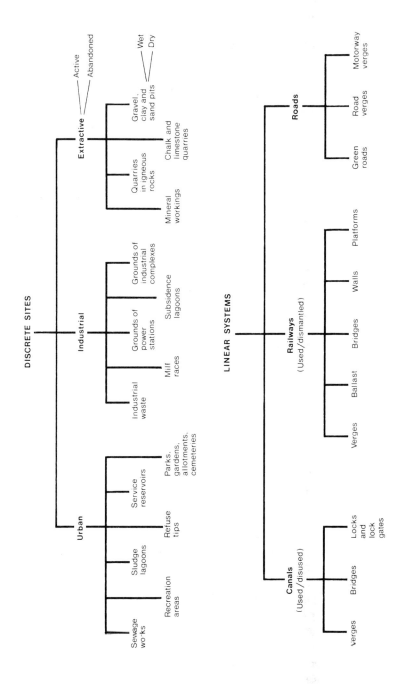

Figure 2.4 Urban and industrial sites of potential interest for nature conservation. (Adapted from various sources (26).)

can thrive in estuaries, due to both the physiological rigours imposed by changing salinities and to the uniformity of the environment, those which can exist often do so in very large numbers because of the abundance of food. This high biological productivity may be of economic importance — directly, for instance in shellfish culture or, indirectly, as food sources for offshore fisheries; human disturbance in such areas may consequently have serious repercussions.

A number of important marginal habitats may also be formed under salt-water conditions, in places where the coast is sheltered in some way; if the water contains an abundant supply of mud and silt in suspension, this may be deposited when stream or tidal velocity is reduced, for instance where the sea overlies a gently shelving, protected coastline or within a wide estuary. In areas which are alternatively covered and uncovered by the tide, only mudbanks are formed, but nearer to the shore tracts of salt marsh tend to develop in the zone covered only by spring high water. Initial colonisation takes place by plants possessing long roots, which in turn help stabilise the shifting sand and mud. Such areas are naturally very fertile, and so have been extensively reclaimed for agriculture; however, the flat nature of the sites, particularly where they are adjacent to a port, has also made them highly attractive to industry. Reclamation practices, have thus led to the destruction of the habitats of many resident species, as well as of migratory wildfowl for which these areas serve as wintering and feeding grounds, often of international importance. The worst effects of this kind of impact can be minimised at the planning stage by careful zoning of uses.

Artificial habitats

Most 'wild' habitats found in Britain today have been modified to a greater or lesser degree by man, and some, such as the Norfolk Broads (which are flooded medieval peat cuttings), are the result of past dereliction. Nevertheless, such sites qualify as 'semi-natural' by virtue of the length of time during which they have been open to colonisation by native species. The large number of habitats created by man's more recent intervention have, however, not been so widely recognised for their ecological importance, and yet this neglect is hardly merited. Broadly, they may be classified as 'discrete' sites, which are mostly of extractive or industrial origin, or as 'linear' systems, primarily comprising transport routes and their associated swathes (26) (Figure 2.4). To these might be added sites more obviously intended for their contribution to urban ecology, such as gardens, parks and landscaping schemes. Urban wasteland and rubbish dumps may also play host to a large number of plants including many well-established aliens; and sewage works have been known to become favoured orni-thological venues.

Whilst sites actively used for industrial or transportation purposes, and therefore generally inaccessible or well protected, may be attractive to wildlife, it is predominantly on disused sites that the greatest potential for colonisation exists. Where good examples can be found, they are of particular importance in the teaching of field studies, as often they are the only areas close to schools which support a reasonable amount of material: the Central Electricity Generating Board, for instance, has already set up field study centres in the grounds of some of its power stations.

Industrial uses have especially benefited the wildlife of lowland Britain by producing sizeable areas of standing water and exposures of basic rocks, two kinds of habitat which are notably rare. In particular, gravel pits, when flooded, have long been known to provide valuable areas for waterfowl and other wildlife, whilst old chalk and limestone workings are likely to support a wide variety of uncommon plants and associated invertebrate fauna.

Reservoirs may mature into lakes of significant biological interest, and in particular many lowland sites provide important wintering grounds for wildfowl, the diversity of species often being greatly enhanced by the varied sequence of vegetation which obtains from the shore into the deepest water, provided this has not been eroded by excessive fluctuations in water level.

Linear habitats are areas upon which communities may establish themselves and channels along which they may spread. Helliwell has recorded 350 unsown species in a 175-mile transect of the M1 motorway whilst Way observes that roadsides are a breeding habitat for 20 of the 50 species of mammal found in Britain as well as all six reptiles, and many birds and insects (27). There has, in recent years, been an encouraging shift by highway authorities away from the use of herbicides and excessively frequent mowing; however, the total abandonment of cutting by some authorities, more in the interests of economy than ecology, is undesirable since it will eventually permit the more delicate herbs to be choked out by aggressive species. The most favourable management regime appears to be a single cut in May or June, except where ground-nesting birds are present when cutting should be avoided before July or August.

It has been estimated that parks, gardens and other open spaces account for 15-20 per cent of the total area of cities and large towns in the UK. These will clearly provide important havens for wildlife, even if their attractiveness is lessened by the widespread practice of planting exotic and ornamental species. One systematic study of a garden in Leicester revealed a variety of butterflies, hoverflies and ichneumonids, which compared favourably with many semi-natural areas, including nature reserves (28). Landscaping schemes similarly provide an opportunity to create refuges within the city, especially where circumstances favour the planting of native species of trees and shrubs.

Creative aquatic landscaping has also been attempted, for instance in the balancing lakes along the River Ouse in conjunction with the expansion of Milton Keynes new town. Among the techniques used were the regulation of minimum depths in order to ensure the survival of fish (about 2·5m), the formation of gently-sloping fringes upon which vegetation could take a hold, the formation of irregularities on the lake beds to provide a variety of micro-habitats, the experimental planting of wild aquatic plants, and the construction of artificial islands as breeding grounds for waterfowl (29).

Plate 8. Road and railway verges provide important corridors along which wildlife may spread.

Site Conservation

It will be apparent from the foregoing descriptions of habitat types that, in a densely populated country such as Britain, there are unlikely to be any genuinely unaltered communities remaining. However, many semi-natural sites, which have been open to colonisation by native species over very long periods of time, are still to be found. As development pressures become more intense, it becomes more desirable to conserve the best remaining sites so that our bank of native genetic material does not suffer irretrievable losses.

The purpose of nature conservation is to sustain natural or semi-natural communities which possess a high degree of integrity. *Conservation* of this type should not be equated with mere *preservation*, for it generally requires careful regulation by man — perhaps even involving the continuation of traditional land management practices. Where it is proposed, therefore, that a site should be solely or primarily used for the conservation of wildlife, it must normally be purchased or leased as a reserve in order that active wardening and management may take place.

Nature reserves range in size from a single field or spinney to thousands of acres. At the top of the scale are the 'biosphere reserves' established by UNESCO, in which an attempt has been made to establish an international network of protected areas encompassing examples of all the world's major vegetation and physiographic types. In addition, they may include: unique communities which contain natural features of exceptional interest, a population of a globally rare species, examples of harmonious landscape resulting from traditional patterns of land use, and examples of modified or degraded ecosystems capable of being restored to more natural conditions (30).

The practical problems involved in the management of a small nature reserve are instanced in the Wharram Quarry reserve, which is under the auspices of the Yorkshire Naturalists' Trust (31). When the quarry was abandoned its floor was being colonised by a wide range of plants including such rarities as the bee orchid and clustered bellflower. Regular visitors to the site, however, were able to discern some deterioration in its ecological quality, and so it was decided to implement a management plan, based on a number of scientific studies. The first of these included a survey of the species present on the reserve — from which compositional changes, in the direction of the likely climax vegetation, floristically less interesting, could be forecast. The different associations, or characteristic groupings, of plants were then identified by a procedure involving statistical sampling and mapping. Those which contained the most interesting species, and those which were approaching the end-stage of the ecological succession could then be distinguished. In those areas where a transition to scrub was apparent, partial clearance of the vegetation was prescribed, thereby allowing the process of succession to be restarted. As the quarry was initially abandoned in 1930, it was assumed, when the studies commenced in 1969, that 40 years was the approximate time taken for the succession to proceed from bare chalk to grassland with scattered hawthorn scrub. It was therefore proposed to repeat this clearance for between an eighth to a tenth of the total reserve area every five years.

The designation of 'key areas' for conservation is, however, inadequate by itself. One aspect of biogeography has been the study of islands, which represent relatively self-contained systems characterised by very restricted inter-migration and invasion, and whose flora and fauna thus display a limited species diversity and genetic variety. What may provide a fascinating avenue of study for the scientist, however, carries a sharp warning for the future of the countryside. As land management practices have become more intensive, with the introduction of highly mechanised farming and forestry systems, so the last remaining fragments of semi-natural vegetation have, effectively, receded to 'islands' within an ecological wilderness. The principles of island biogeography, therefore, tell us that these retreats will experience declining inter-migration from the denizens of similar habitats, as well as adverse 'edge effects', whereby species alien to the community start to invade. These impacts are made worse by the widespread removal of interconnecting linear habitats, such as hedges, drainage channels and road verges.

As a consequence, conservationists are turning their attention towards the maintenance of ecological diversity in the countryside at large: there is little point in protecting, say, a rare species of predatory bird on a reserve if its prey has been eliminated from the surrounding areas. Since planners are, in the broadest terms, concerned with land use change, they must become centrally involved in conservation strategies, seeking to maintain as wide a range of habitats as possible, even where agriculture, forestry or some other activity is the primary user (32).

Ecological Dynamics and Regional Planning

There has been a discernible shift in modern ecology away from the study of spatial variation in the form of ecosystems to their dynamic functioning in terms of flows of energy and cycles of nutrients. The ultimate source of energy is the sun, and this can be tapped directly by green plants and by certain bacteria and algae through a process known as photosynthesis by which organic matter is manufactured from water and carbon dioxide. This food energy locked up in plants is then consumed by herbivores who are in turn preyed upon by carnivores; thus energy flows along a 'food chain'. In practice, this simple hierarchical feeding relationship rarely exists and the complex organisations which are observed in most ecosystems are more aptly termed 'food webs'. The filigree of linkages within a food web is so intricate that the removal of one organism — for instance, by land reclamation or pollution — may have quite unforeseen effects on other species which were dependent upon it at some stage in their life cycles.

Such is the complexity of these webs that it is usual to generalise ecosystem structure in terms of the abundance of organisms at each feeding, or 'trophic', level: thus plants (primary producers) represent the first level, herbivores (primary consumers) the second, carnivores (secondary consumers) the third, and top carnivores (tertiary consumers) the fourth. The transfer of energy between each level is rather inefficient, often around ten per cent (33), and it follows that the larger animals, of the higher trophic levels, will be relatively rare. This hierarchy may be represented graphically as a 'pyramid of numbers' with each tier, of which there are rarely more than four, representing a specific trophic level (Figure 2.5).

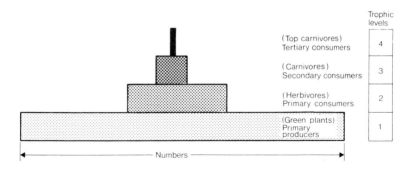

Figure 2.5   Pyramid of numbers.

Energy, although it cannot be destroyed, is ultimately dissipated from an ecosystem, and its flow is thus uni-directional. Minerals, which furnish the essential nuitrients for life, conversely, are retained in the system and are constantly being recycled; thus the decay of dead animals, plants and leaves, under the action of decomposer organisms, returns minerals to the soil, whilst nutrients are also being added by the weathering of underlying parent material and by the fixation of nitrogen from the atmosphere by certain bacteria (Fig. 2.6). These 'biogeochemical cycles', especially those of the three major nutrients — nitrogen, phosphorus and potassium — are essential to the maintenance of the productivity of ecosystems and are fundamentally dependent upon the abundance of soil micro- and meso-fauna. It has been contended by a number of critics that certain modern land management practices, such as continuous cropping and the application of heavy dressings of artificial fertilisers, may be permanently debilitating the health of the system; however, in most cases, the evidence is inconclusive.

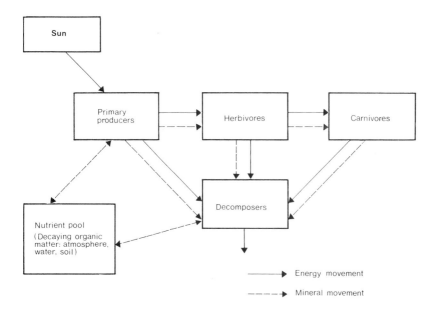

Figure 2.6 Undirectional flow of energy and cycling of nutrients in an ecosystem.

It is clear, however, that both the biotic and abiotic components of our natural environment, resilient though they may be, possess limits beyond which exploitation might cause irreversible damage. It follows that, if human communities are to remain in a state of dynamic equilibrium, efforts must be made to contain their ecological demand within the carrying capacity of their environment. Thus, for instance, the State Government of Hawaii has sought to establish in its legislature '... criteria for defining the state's optimum carrying capacity as related to its ... population, air quality, water quality and supply, energy supplies, transportation systems and land use capabilities' (34). However, a separate study by Ricci, in which an attempt was made to define a national carrying capacity for Canada, showed just how difficult it is to define a single optimum level for human communities, so sensitive did it prove to institutional and political pressures and to technological innovation (35).

In general, for natural communities in a steady state, the following balance equation holds true:

$$I \quad + \quad P_e \quad = \quad D \quad + \quad L$$

(import of energy and materials from outside ecosystem)   (net production of ecosystem)   (decomposition)   (losses of energy and materials).

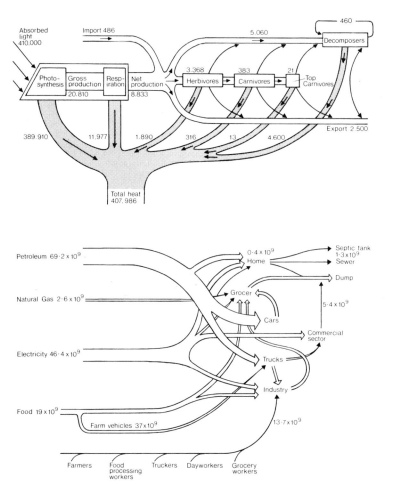

Figure 2.7 Energy budgets showing food consumption patterns in two communities. Above: *A natural community, Silver Springs, Florida (units = kcal/m¹/yr).(Source: Odum (33)).* Below: *A human community – the State of Oregon (units = kJ/yr). (Source: The State of Oregon, Special Projects Branch (36).)*

With regard to human communities, imports may be equated with resource subsidies, net production with industrial and biological capacity, and decomposition and losses with the wastage of energy and materials (ie pollution). In the case of natural ecosystems, ecologists have succeeded in constructing 'energy budgets' which quantify imports and production; some pioneering attempts have been made to extend this modelling technique to human communities, and these assist greatly in clarifying the nature of our use of resources (36) (Figure 2.7).

In advanced societies, the levels of subsidy required will be enormous: for instance, whilst the least developed countries import energy subsidies of only about 3-4 000 kcal/day/capita, the equivalent figure for the USA is some 220 000. It will be obvious that, unless strict preventive measures are taken, commensurate amounts of pollution will be produced. Two key aspects of the ecology of human communities thus concern the planner: the energy, mineral and other resources of our abiotic system (Chapter 1) the impacts on our environment consequent upon the exploitation of these resources (Chapter 3).

NOTES

(1) F. B. Goldsmith (1974) 'Ecological effects of visitors in the countryside' in A. Warren and F. B. Goldsmith (Eds.) *Conservation in Practice*, London, Wiley.

(2) F. H. Perring (1967) 'Changes in chalk grassland caused by galloping' in E. Duffey (Ed.) *The Biotic Effects of Public Pressure on the Environment*, Nature Conservancy Monks Wood Symposium No 3; this research is reviewed in M. B. Usher (1973) *Biological Management and Conservation*, London, Chapman and Hall, pp. 279-281.

(3) J. Newton and M. Pugh Thomas (1979) 'The effects of trampling on the soil Acari and Collembola of a heathland', *International Journal of Environmental Studies*, 13.

(4) I. Richards and F. Stead (1977) 'Dune conservation in Anglesey', *Town and Country Planning*, 45.

(5) Countryside Recreation Research Advisory Group (1970) *Countryside Recreation Glossary*, Cheltenham, Countryside Commission.

(6) R. G. H. Bruce (1976) 'Native Pinewoods of Scotland', *Scottish Wildlife*, 11.

(7) C. R. Tubbs (1974) 'Woodlands: their history and conservation' in A. Warren and F. B. Goldsmith (Eds.) *Conservation in Practice*, London, Wiley.

(8) O. Rackham (1976) *Trees and Woodland in the British Landscape*, London, Dent.

(9) Joan Davidson and G. Wibberley (1977) *Planning and the Rural Environment*, Oxford, Pergamon.

(10) D. I. Brotherton (1977) 'Lowland grasslands' in Joan Davidson and R. Lloyd (Eds.) *Conservation and Agriculture*, Chichester, Wiley.

(11) This topic is further discussed by E. Duffey (1974) *Grassland Ecology and Wildlife Management*, Institute of Terrestrial Ecology, Monks Wood Experimental Station; and E. Duffey (1974) 'Lowland grassland and scrub: management for wildlife' in A. Warren and F. B. Goldsmith, *op. cit.* (1).

(12) W. H. Pearsall (1950) *Mountains and Moorlands*, London, Collins.
(13) J. B. McCreath (1976) 'Land reclamation in the hills and uplands of Scotland' in J. Lenihan and W. W. Fletcher (Eds.) *Reclamation*, Glasgow, Blackie.
(14) Lord Porchester (1977) *A Study of Exmoor*, London, HMSO.
(15) A seminal discussion of the ecological effects of modern farming practices was provided by R. K. Cornwallis (1969) 'Farming and wildlife conservation in England and Wales', *Biol. Conserv.*, 1.
(16) R. Westmacott and T. Worthington (1974) *New Agricultural Landscapes*, Cheltenham, Countryside Commission.
(17) Council for the Protection of Rural England (1975) *Landscape – the need for a Public Voice*, London, CPRE.
(18) Heathland ecology is discussed by C. H. Gimmingham (1974) *An Introduction to Heathland Ecology*, Edinburgh, Oliver and Boyd; and by G. R. Miller and A. Watson (1974), 'Heather moorland: a man-made ecosystem' in A. Warren and F. B. Goldsmith, *op. cit.* (1).
(19) N. W. Moore (1962) 'The heaths of Dorset and their conservation', *Journal of Ecology*, 50
(20) Carolyn Harrison (1974) 'The ecology of British lowland heaths' in A. Warren and F. B. Goldsmith (Eds.), *op. cit.* (1).
(21) S. M. Haslam (1973) 'The management of British wetlands', Pts I & II, *Journal of Environmental Management*, 4.
(22) A. R. Hill (1976) 'The environmental impact of agriculture land drainage', *Journal of Environmental Management*, 4.
(23) C. K. Catchpole and C. F. Tydeman (1975) 'Gravel pits as new wetland habitats for the conservation of breeding bird communities', *Biological Conservation*, 18.
(24) See, for instance, D. H. Mills (1972) *An Introduction to Freshwater Ecology*, Edinburgh, Oliver and Boyd.
(25) R. S. K. Barnes (1974) *Estuarine Biology*, London, Edward Arnold.
(26) J. G. Kelcey (1975) 'Industrial development and wildlife conservation', *Environmental Conservation*, 2; B. N. K. Davis (1976) 'Wildlife, urbanisation and industry', *Biological Conservation*, 10; M. B. Usher (1979) 'Natural communities of plants and animals in disused quarries', *Journal of Environmental Management*, 6; A. Bradshaw (1979) 'Derelict land — is the tidying up going too far?', *Planner*, 65.
(27) D. R. Helliwell (1974) 'The value of vegetation for conservation' Pt II M1 Motorway area', *Journal of Environmental Management*, 2; J. M. Way (1977) 'Roadside verges and conservation in Britain: a review', *Biological Conservation*, 12.
(28) Jennifer Owen and D. F. Owen (1975) 'Suburban gardens: England's most important nature reserve?', *Environmental Conservation*, 2.
(29) J. G. Kelcey (1978) 'Creative ecology: Pt 2, Selected aquatic habitats', *Landscape Design*, 121.
(30) O. Frankel (1978) 'Biosphere reserves: the philosophy of conservation' in J. G. Hawkes (Ed.), *Conservation and Agriculture*, Duckworth, London.
(31) M. B. Usher (1976) *Biological Management and Conservation*, London, Chapman and Hall, pp. 321-244.
(32) P. H. Selman (1976) 'Wildlife Conservation in structure plans', *Journal of Environmental Management*, 4.
(33) J. Phillipson (1966) *Ecological Energetics*, London, Edward Arnold. See also H. T. Odum (1957) 'Trophic structure and productivity of Silver Springs, Florida', *Ecological Monographs*, 27.
(34) D. R. Godschalf and F. H. Palmer (1975) 'Carrying capacity — a key to environmental planning?', *Journal of Soil and Water Conservation*, 3.

(35) P. F. Ricci (1978) 'Policy analysis through carrying capacity', *Journal of Environmental Management*, 6.
(36) State of Oregon, Special Projects Branch (1973) *Energy and State Government*, Office of the Governor, Salem.

3 Competition for Resources

Agricultural and industrial revolutions have enabled man dramatically to increase the size of his numbers beyond the apparent carrying capacity of his environment. In part, this has been brought about by the development of specialised crop strains and the consequent management of agricultural monocultures. Within the past two centuries or so, however, the separation of man from his natural constraints has been greatly accelerated by the importation of huge subsidies of that most vital of species requirements: energy. This in turn has unleashed unprecedented rates of population growth, which have brought in their wake demands for subsidies of other essential resources, such as land, water and leisure space (Figure 3.1).

The apparent success with which man has conquered the elements has given an illusion of independence from nature which has proved both unrealistic and dangerous. This chapter therefore surveys some of the more marked impacts which the subsidy requirements of urban man engender, an approach which is intended to provide a rationale for the inclusion of a broad range of 'environmental' issues under the more specific heading of 'ecology'. Elsewhere the other aspect of human resource systems evident from Figure 3.1 is reviewed: that of pollution. The subsystems within this framework all suggest density dependent and independent constraints which would spell destruction for most other species displaying such voracity and rampant growth, but in the case of *Homo sapiens* these limits have proved remarkably flexible and obliging. This is largely attributable to the high degree of environmental resistance (*vid.* Figure 0.2) provided by our economic system, which modifies our perception of resources and suggests alternative strategies for their development and conservation.

To many observers, however, this propitious relationship, this delicate balance, is being upset by political tensions, rates of resource consumption and sheer pressure of numbers: it is as if time had started to run out on human inventiveness. Fortunately, man does not have to accept the inevitability of impending limits, for his ability to plan for

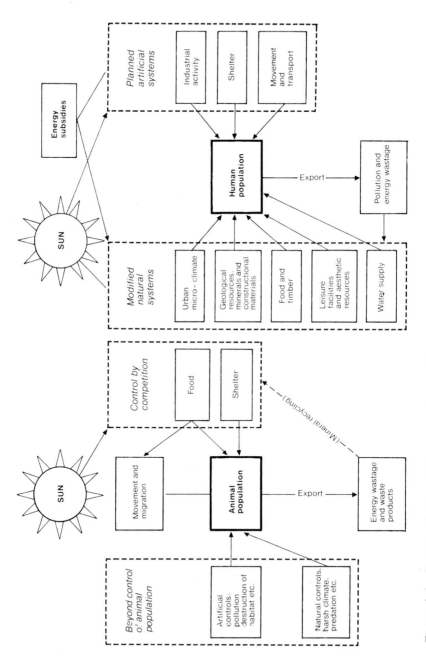

Figure 3.1 Limiting factors in human and natural ecosystems.

the future, to adapt, and to respond constructively to deteriorating conditions sets him apart from other animals. It is contended that such a response is only possible within an overall appreciation of the ecology of human communities.

What is a Resource?

Before considering the nature of, and impacts consequential upon, competition for natural resources, it will be useful briefly to consider what a resource actually is. A resource has been defined as an element of the natural environment appraised by man to be of value, but whose supply falls short of demand; thus, it is as much a cultural concept as a physical entity. In the first instance, therefore, a specific use must have been identified: before its discovery as a potent fuel source, for instance, uranium was considered to have only a small value as a green pigment for glass. Secondly, the supply of the element must fall short of our present, or perceived future, requirements. Broadly speaking, resources may be categorised according to their renewability or non-renewability, the former becoming scarce through exploitation beyond the rates of natural replenishment (eg an ocean fishery, or a recreational area in which vegetation is becoming eroded by excessive pressure), and the latter through exhaustion of a finite stock (eg a petroleum well). Some non-renewable resources may, however, be recycled, provided they have not become too degraded during their usage.

Thus, 'stock energy resources', in which re-cycling is effectively impossible, comprise inventoried past receipts of solar energy — fossil fuels — and have a fixed and finite stock available for depletion. 'Stock material resources', such as metal ore deposits, also comprise a finite stock, but re-cycling is possible. 'Flow energy resources' include continuous natural energy sources, such as solar, tidal and wind power, for no amount of current consumption can reduce future availability. Finally, 'stock renewable resources' consist of animal and plant populations, and are characterised by having, at any point in time, a fixed stock size which is renewable. Within the human economy, therefore, the throughput of resources necessary for a given level of production decreases as efficiency of energy conversion, materials usage and re-cycling increases. In this context, it is important to appreciate the concept of a 'materials balance', which states that inputs of useful resources into a system must equal the outputs from it (ie pollution, or resources 'out of place'); it is clearly one of the tasks of planning to seek to reduce these inputs and outputs to minimum levels which are compatible with continued economic well-being.

Urban Encroachment

Urbanisation is an inevitable concomitant of economic expansion and is the single greatest usurper of the countryside, its resources and landscape. As the human race continues to increase its numbers, so towns and cities continue to multiply and expand: the more land used for human shelter the less there is for food production.

In Britain the 1947 Town and Country Planning Act was largely a response to the rapid and wasteful sprawl of urban areas which occurred between the wars (1918-1939). It ushered in a planning system which, although admirable in its way, is now seen in retrospect to have been less effective in the matter of protecting the countryside than had been assumed. The Director of the Second Land Use Survey of England and Wales, Dr Alice Coleman, decided to re-survey certain areas in order to assess the rate of land transfer (1), and produced results which suggested that the rates of transfer of agricultural land to urban use were frequently as high as those prior to planning legislation; indeed, many such losses seemed to be the result of the planning system itself. In particular, land was often inefficiently used — being neither truly urban nor truly rural (the Coleman term was 'rurban'). There was therefore a case for a more intensive use of land on the basis of its intrinsic capability: the conversion of indeterminate tracts to either 'townscape', 'farmscape' or 'wildscape' would not only ensure a more efficient use of land stock but would also be aesthetically satisfying.

Coleman's critics, however, argue that many of the areas which were re-surveyed — the London Docklands and Berkshire suburbia, for instance — were atypical and that any general conclusions about rates of land loss must therefore be treated with caution. Alternative figures suggest that the average rate of land transfer subsequent to the 1947 Act has been rather less than two-thirds that of the 1930s (Figure 3.2).

Although it is difficult to ascertain the rate and extent of urban growth with certainty, it appears that only ten per cent of the land surface of England and Wales was under urban use in 1961 and that it is unlikely to increase beyond 14 per cent by the end of the century (2). However, the continued conversion of some 13,000 to 20,000ha annually in England and Wales and some 2,700ha in Scotland has alerted many planning authorities to the need to make the recycling of vacant inner city land a matter of planning priority and to give more careful consideration to the agricultural worth of 'greenfield sites'.

Whilst the actual area covered by urban uses has been, on the whole, effectively controlled as a result of planning legislation, the indirect pressures of towns and cities have spread well beyond their peripheries the result being the indeterminate pattern of land use which Coleman

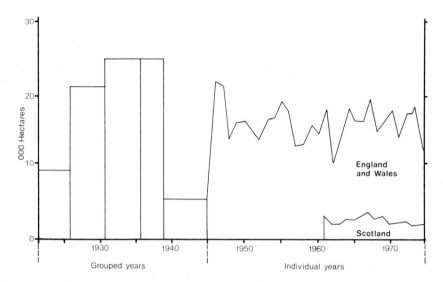

Figure 3.2 Land transfers from agricultural use, 1920-1975. (Source: Best (2).)

has identified. Many of these urban fringe areas were zoned as 'Green Belts' — a designation initially intended to contain cities and to retain rural amenities nearby. Now, many have become little more than receptors for a variety of urban uses — sanatoria, service reservoirs, sewage works. Often they are prime settings for conflicts between individuals/agencies seeking fairly exclusive use of their sites and the general public who expect widespread access on an informal basis (3).

More positive management is required for the Green Belt areas. Local authorities should actively promote conditions amenable to the continuation of viable farming. For example, access agreements with landowners could be entered into so that vandalism and trespass, endemic in these peri-urban places, are reduced; the re-orientation of farming enterprises could be encouraged to take positive advantage of local demand for produce. The longer term improvements are many: radical restructuring of landscapes, re-shaping of farms, rationalising road and footpath systems, restoring derelict land, regenerating wood-land and providing attractive and accessible leisure facilities.

Energy

Coal

Coal mining and its ancillary industries have left a legacy of pit-head machinery, waste tips, coal storage sites and subsidence largely taken for granted by traditional mining communities. Even the disaster at Aberfan in Wales only promoted a limited response in ensuring the greater stability of waste tips and, in some cases, their rehabilitation. To some extent, this apparent apathy could be attributed to the assumption that the coal industry was passing into history, and that its ravages would gradually be healed by time.

The 1970s, however, brought the realisation that, whilst economic oil deposits might only have a life of 50 or 60 years, Britain possessed a minimum of 300 years' recoverable coal reserves which would play an integral role in any future energy strategy. The discoveries of massive fields at Selby and the Vale of Belvoir re-inforced the message that coal mining was here to stay. In addition to the complexes of power stations which tend to accompany coal, new mining communities may also have novel technological processes, such as coal conversion plant, bestowed on them. The new coal deposits of the Midlands are no longer to be found in the scenically rather bleak and historically despoiled tracts of Carboniferous countryside, but under the visually and agriculturally more valued Triassic deposits of the 'hunting shires'. Further, these strata tend to be water-bearing, and so in order to reach the Carboniferous deposits beneath them, cased 'shaft mines' must be sunk, visually far more intrusive than the 'drift mines' which are often feasible elsewhere. Not surprisingly, the public inquiry into the proposed Vale of Belvoir coalfield became one of the environmental *causes célèbres* of the 1970s (4).

Where coal is extracted on an opencast basis its impact, although substantial, is largely temporary; in the case of deep mining, however, disruption can be long-term (5). Land lost to pit-head and associated facilities has been increasing over time, and can be considered a permanent loss. The associated requirements for waste disposal and coalstocking have also shown large increases, and although reclamation is possible, the original quality cannot be regained, despite some recent successes in re-establishing arable crops on reclaimed land. Subsidence may affect both buildings and agricultural land drainage systems; although its impact is relatively localised and predictable, and compensation may be paid, it is still perhaps the phenomenon which causes the greatest apprehension amongst affected communities. Apart from the visual intrusion of the plant itself, its effects of air pollution, arising from coal preparation and spoil heaps, and of noise pollution, associated with plant construction and operation, traffic and tipping, may cause a more serious disamenity to local residents, whilst polluted water, especially from spoil tip run-off, has been known to reduce seriously the quality of adjacent streams.

Despite these problems, coal is immensely important to Britain's future, and there is a considerable onus upon planners to ensure that it is recovered with the least possible damage to the environment, whilst ensuring that the development of mining and its associated industries is coordinated in the most beneficial manner to the economy. This task should be made easier with the recent statement of intent from the Department of the Environment that more of the National Coal Board's operations will be brought under planning control.

Oil and gas

But the discovery of oil and gas reserves in the North Sea have, of course, had the greatest impact on Britain in recent years. The environmental problems of their exploitation off the Scottish coast, given that Scotland contains almost half of the total UK area of conservation interest, are unprecedented (6). The offshore oil industry pioneered its first large-scale ventures into waters deeper than 100m; this increased both the hazards of operation and the scale of onshore ancillary activities. Conditions in the North Sea have proved to be some of the worst ever encountered: high winds, large waves with complex wave patterns combine to disrupt the operation of survey and supply vessels; the installation of production platforms, pipelines and of tanker loading facilities greatly increases the risks of pollution. Related industrial development, notably platform and module construction, was at first confined to the east coast, at Nigg and Ardesier for example, so that structures could be erected as near as possible to where they were to be positioned. However, the subsequent change to the use of concrete platforms required very deep water sites in which the upper structures could be added to the lower section by slipforming as the latter was lowered progressively into the sea. Such depths were available only along the unspoilt coast of the West Highlands.

The challenges to planning authorities in Scotland were many and various; in the words of one early commentator '... For the hard pressed authorities who are dealing with applications without plans made in the knowledge of the effect of oil discoveries, and without coherent criteria by which to judge them, there is the further problem of guiding development so that in twenty-five years time the remains of North Sea oil exploration will not scar what is now the only large area of "wilderness" in the UK...' (7). The Scottish Development Department published, among various other general advisory and information papers, Coastal Planning Guidelines designed to steer on-shore servicing and constructional installations into 'preferred development zones' thereby avoiding 'preferred conservation zones' where there is strong presumption against oil-related development (Figure 3.3) (8). More recently the Scottish Development Department has been encouraging local authorities to prepare contingency plans for the location of possible petrochemical processing plants for the consumer-

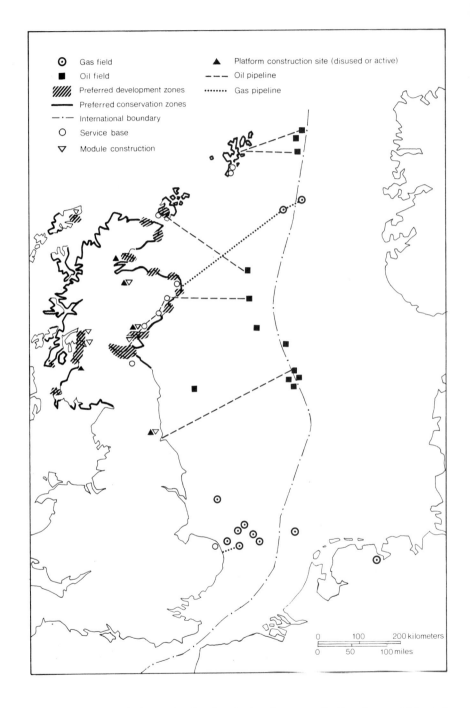

Figure 3.3   North Sea oil – the developing impact of oil and gas. (Adapted from Smith et al *(6), SDD (8).)*

orientated industries about to embark on another phase of the oil boom. Once again the effort is being made to implement plans which will allow the greatest long term benefits whilst giving the maximum protection to the environment.

Nuclear energy

In many respects, the civil nuclear programme offers the best alternative were it not for the implications of potential local radiation hazards and an international trade in plutonium. At the public inquiry held into the proposed reprocessing plant at Windscale, Cumbria, many of the considerations which the Inspector had to consider were not materially different from those associated with a great many other projects — however the safety aspects clearly differed enormously, both in kind and in scale (9). On the face of it, fast breeder reactors accord with the best principles of re-cycling of resources: by breeding new fissile material from their uranium and thorium during irradiation, they produce more plutonium than they are able to consume; this can be recovered, re-irradiated and repeatedly re-used. But this cycle is not totally efficient and results in the creation of a number of waste products many of which will emit powerful amounts of radiation into the lifetimes of many future generations.

Nuclear power generation gives rise to wastes which may be liquid (of high, medium or low activity), solid (of high or low activity) or gaseous. If their activity is low enough, they may be released to the environment, but if high they must be stored, either indefinitely or until the activity has decreased to the point where control over them can be relaxed. Combinations of these alternatives may at times be possible: for instance, a low or moderate activity waste may be treated to give partially purified waste suitable for immediate discharge and a sludge which must be stored for eventual decay and release.

Two types of by-product are associated with nuclear waste, fission products and actinides, the former generating intense radiation, but having half-lives of only around thirty years and thus becoming harmless relatively rapidly, and the latter being hazardous to man if directly ingested or inhaled, but containing species with half-lives of hundreds or thousands of years. The safe disposal of wastes thus requires a very deep repository which will not require continuing human surveillance and which is stable geologically (ie unlikely to experience tectonic disturbance). The major land use planning implication of this is that applications for waste disposal are likely to be confined to a relatively few locations which possess suitable geological formations. Residents living in such areas have often responded with alarm, claiming that the behaviour of vitrified wastes, and of the metal containers in which they are cased, over long periods of time and under conditions of deep burial, is largely unknown. In Cheshire, for instance, concern has been expressed about the suitability of the hybrid clay and salt formations which have been suggested for waste disposal. It has been pointed out that small inclusions of brine within the salt might migrate towards the waste containers and produce a very corrosive environment; with regard to the clay deposits, little is known about their structural properties at depth or response to any heating which might be associated with radioactive waste.

Whilst dumping requirements arising out of nuclear power are planned and predictable, there also exists the possibility, admittedly very remote, of a nuclear accident. Very great care is taken in the design, construction and operation of reactors to make them accident proof; with the exception of the accident at Harrisburg, USA, in 1979, there have been, apparently, only minor occurences. Every reactor installation must, nevertheless, have a detailed emergency plan adequate to protect individuals in the vicinity since there is still a risk of the fuel overheating as a result of mechanical failure in the coolant circuit and this, it seems, could cause an accident serious enough to release fission products to the environment. There is also the possibility, although even more remote, that the core might suffer far greater damage and secondary containment devices might simultaneously fail: then the emergency plan would be overwhelmed and radiation exposures could occur at distances of up to thirty or forty miles.

Electricity

Electric power generation imposes its own constraints on the environment. Distribution is by national grid and there is the consequent need to feed in electricity at key points within the network rather than at sites adjacent to centres of consumption. Modern coal-fired and nuclear stations usually tend to be sited at coastal locations to avail of the marine access and the abundance of cooling water, but some coal fired stations are still sited inland close to mining areas requiring large complexes of cooling towers to recycle the limited quantity of water. These are intrusive on the landscape and there is generally the risk of thermal pollution of associated rivers. Hydro-electric stations also prove controversial since their scale and character may conflict with their relatively undisturbed upland settings. Recently, a number of pumped storage schemes have been established in which off-peak electricity is used to pump water from a lower to an upper reservoir, the latter providing conventional hydro-electrical generation during peak hours (10). The large volume of water required in the lower lake and the substantial height differential needed between this and the upper reservoir have narrowed the choice of suitable sites to a few

Plate 9.   Energy developments almost invariably occasion major impacts on the built and/or natural environment.

remote areas of scenic grandeur. The pumping regimes tend partially to disrupt the thermal stratification and ecological processes of the lower lakes, some of which are of major biological significance.

Flow energy resources

Investigation into alternative methods of electricity generation, especially those based on 'flow energy resources' (sun, tides, waves etc), continues and there are those who claim that in the long term these will prove both viable and preferable. Such sources are considered 'low-impact' in relation to their demand on finite reserves; yet, even they could give rise to fairly contentious planning issues. For example: the land-take requirements of hydro-electric power, the disruption of shipping lanes by tidal and wave power, the inundation of wildlife communities by estuarine barrages, giant windmills ('aero-generators'), the replacement of traditional materials by solar panels could all prove controversial.

Future Solutions

Increasing impact upon the natural environment must be expected if requirements for energy subsidy continue to rise. The degree of impact will vary according to the relative importance of each fuel in future energy programmes (11). Widely differing estimates for energy production and consumption in the future are based on such premises as whether we seek to continue present patterns of usage, whether we seek to reduce rates of economic growth, or whether we maintain growth trajectories by maximising efficiency of fuel and power production and conservation — known as the 'technical fix solution'.

The forecasting of future energy requirements is particularly problematic, since there is a high degree of potential substitution between fuels, particularly in the long term, so that the actual 'mix' of energy sources is difficult to forecast, whilst there is also major uncertainty with respect to conditions of supply and demand. Nevertheless, planners must make adequate provision for energy supply, since investments have very long lead times, and demand for fuel, especially in the short term, is very inelastic, rendering considerations of security and continuity of supply dominant. Energy forecasts for Britain have been repeatedly revised downwards since 1973, reflecting the combined effects of economic recession and fuel conservation policies; current official predictions foresee a primary fuel requirement of 445-510m. tonnes coal equivalent (mtce) in 2000, representing an average annual growth rate of 0·5-0·9 per cent a year, although an independent study has suggested that, given the systematic application of known conservation measures, Britain could have zero energy and electricity growth for the next 50 years, even if GDP is trebled.

Given this degree of uncertainty, it is difficult to provide clear future guidelines for the land use implications of energy development. The Watt Committee on Energy, which reported in 1979, however, felt that major future impacts could be experienced in Yorkshire and the Midlands (coal), the west coast (potential oil and gas discoveries), main inland rivers and estuaries and certain coastal locations (thermal and nuclear power stations), north and west Scotland (wind and wave resources) and the Severn and Solway Firths (tidal barrages). A particular problem was felt to be the possible shortfall of suitable nuclear power station sites beyond the end of the century, when near-urban sites could become necessary.

Minerals

Mineral extraction is one of the few economic activities which is absolutely constrained in terms of location: minerals can be worked only where an economic deposit occurs, and this is as likely to be in a National Park as in an area of limited amenity. From the planner's point of view, economic factors such as the value of the mineral per unit of material extracted, the value added by on-site processing and the volume of production needed to make operations economic, are often of greater significance than other considerations.

Certain minerals, such as the major aggregates for road-building and construction, have a high 'place' value — or low worth in relation to transport costs — and increase in value only slightly as a result of on-site treatment. Consequently, they tend to be worked in close proximity to their markets, so that although they may prove a considerable disamenity when active, a great deal of scope exists for the creative after-use of abandoned sites; conditions of planning permissions nowadays often include restoration to recreation and conservation use.

Other minerals, such as chalk, limestone, brick clay and rock salt may not in their raw state be particularly valuable, but gain in value substantially from on-site processing and may be marketed nationally. Finally, certain relatively valuable minerals and ores occur in the UK, the working of which will contribute significantly to the balance of payments and enhance local employment prospects, and they are therefore likely to be worked even in the face of strong opposition. They include the chief raw materials of the chemical and fertiliser industries such as potash (which was the subject of a celebrated environmental controversy in the North York Moors), the relatively high value non-metallics, for instance china clay and fluorspar, and metalliferous ores which, because of their relatively low grade in Britain, are often worked on an extensive opencast basis (12).

Water

Water supply represents the major density dependent factor affecting most human populations. In Britain the main problem is not so much one of water shortage as one of rainfall distribution — most of the rain falls in the north-west and most of the population live in the south-east. The traditional solution to this has been the impoundment of water in upland areas, an approach which has become increasingly controversial because of its visual impact and because of its disruption of local agriculture. The construction of reservoirs has given rise also to considerable ecological controversy, for instance at Cow Green in Upper Teesdale, in which part of an assemblage of arctic-alpine flora was to be inundated, with an attendant possibility that the slight microclimatic improvement induced by the reservoir might endanger the remainder of the community (13).

Problems may be even more acute where major new reservoirs are proposed in lowland areas, although the opportunities for multiple use are often much greater, and the leisure needs of the relatively nearby centres of population are generally fully considered at the design stage.

These environmental and economic costs of impoundment are, however, now beginning to militate in favour of alternative sources, such as groundwater abstraction and the transfer of water between drainage basins, which raises strong arguments in favour of reducing pollution levels in natural watercourses. In the latter method, rivers, supplemented where necessary by aqueducts, are used to transfer from one catchment area to another (14). It has been suggested that this practice might disrupt salmon migration patterns in some of the affected rivers, notably the Usk and Wye, since the resultant mixture of river water could prove unrecognisable to returning fish (15).

It is estimated that, on the basis of current capacity, local shortfalls in supply could rise to $12 \cdot 1$ m metres^3 per day by the year 2001. Only a part of these requirements can be met from conventional sources, even allowing for higher levels of recycling. If we are to continue to meet demand, then large-scale schemes with extensive ecological repercussions, such as the construction of estuarine barrages, will have to be considered. The alternative approach, which has been adopted in some countries, is to reduce demand by metering water consumption (generally leading to a reduction in domestic consumption of about nine per cent), and to supply water of a lower quality to agriculture and industry.

Transport

The linear patterns of transport networks affect air land and water systems, intersecting communities, both human and wild, transforming the quality of life in human settlements and scenic/ecological features in landscape. The growth of the railways in the 19th century spread urbanisation over the countryside in a manner and tempo never before envisaged. From the vantage point of the late 20th century, the effects of steam locomotion on the built and natural environments are seen as having been relatively mild, even benign; railway architecture and old rolling stock are now cherished as part of the national heritage — and railway embankments have long supported a wealth of wildlife.

The motor car has perhaps been the major factor in disseminating urban pressures into the countryside, and major road proposals have produced some of the most bitter environmental confrontations. The road users' lobby is understandably vociferous and single-minded: to many people, a car is their major expression of individual freedom, and is certainly a singularly expensive one. Their opposition to restrictions on road construction, improvement and usage is therefore perfectly comprehensible. However, whilst nothing can mitigate the enormous land loss inevitably associated with modern highways, much can be achieved at the corridor selection and detail design stages to soften their impact on the landscape and wildlife (16).

Air traffic too creates difficulties. Even relatively minor extensions to provincial airports are vehemently opposed by local residents — as recent experience at Edinburgh and Birmingham attests — whilst the proposal for a third London airport raised issues of such magnitude that its siting became the subject of a Royal Commission (17). Residents at Stansted mobilised powerful opposition to the Commission's recommendation that their peaceful village should become the venue, whilst a dissenting member, in a minority report, proposed Foulness ('Maplin') as being least detrimental to the total environment. This latter view was accepted by the Government of the day, although current preferences seem once again to be for Stansted. Even the remote site at Maplin Sands, however, proved controversial: it was the major wintering ground in Britain for the Brent Goose, whose only food (it was then supposed) was the eelgrass found in abundance on the sands. Birds can take their own revenge, however; they pose hazards of bird strike to aircraft (18).

Air pollution and noise

Apart from the visually intrusive effects of motorways and airports on a landscape, their pollutant nature has made them the target of wider public opprobrium.

In respect of air traffic, the degree of annoyance associated with noise is related both to the total number of aircraft heard and the average peak noise level as each aircraft approaches and recedes from the observer. These parameters are combined in the Noise and Number Index (NNI) and contour maps may be prepared for areas around airports to show the incidence of critical NNI levels. For example, the Swiss authorities, have attached particular significance to the 55 NNI contour around Zurich airport; further residential development has been prohibited within this contour and special protection against noise has been provided for housing within the 45-55 NNI zone (19).

Road traffic noises cannot be quantified in the same way. Instead, an index has been devised to show the highest noise level, expressed on a logarithmic scale, which is exceeded for ten per cent of the time during a specified 18 hour period: the L_{10} (18h) Index. The scale chosen is the 'A-weighted' decibel scale — dB (A) — and it is currently recommended in Britain that the L_{10} level outside a house should not exceed 68dB (A). Efforts at minimising noise include the construction of barriers around dwellings, and screening with trees and shrubs to camouflage the source of disturbance can achieve beneficial phsychological effects.

Air pollution is associated particularly with the emission of carbon dioxide and lead in exhaust fumes. But there are some remedial actions which can be taken — measures aimed at smoothing the flow of traffic, for example, will reduce the amount of carbon monoxide emissions. Restrictive action of the lead content of petrol is the prerogative of central government and has not, as yet, been taken in Britain*; but in north America and some European countries lead is no longer permitted in transport fuels. Some disturbing claims have been made about the effect of lead pollution: studies carried out in Boston, New York, Montreal and Düsseldorf confirm the sub-clinical effects of lead on children's mental performance; a study on the milk teeth of children in Birmingham has revealed that the majority of the city's children have enough lead in their bodies to warrant deeper public concern. A Government Working Party on lead pollution at a site near the Gravelly Hill motorway interchange ('Spaghetti Junction') concluded that the amount of lead in children living in this area was not medically significant, although a dissenting member of the team suggested that perhaps twenty per cent of children in the city's inner area, under the age of thirteen, are actually experiencing a disturbance of central nervous system functions (20).

High lead levels have also been found on roadside vegetation and in small mammals. This is clearly an important access point where toxins can enter ecological pathways. These specimens were found along routes with heavy traffic flows, the M1 Motorway and part of the A1 but it has been shown that lead levels decrease rapidly with distance from source (21).

Single purpose transport media

Pipelines, overhead transmission cables and other single purpose transportation media are increasingly being extended throughout the countryside. Whilst careful route selection and landscaping may mitigate the visual effects of electricity lines, little can be done about the restrictions on land use under their path, or the uptake of ground by pylons. Those sited underground, although not visually intrusive, nevertheless impose restrictions on land use since wayleaves for repair and maintenance are required; also, vegetation along such routes may suffer scarring and this may be quite severe and extensive given the considerable width of swathes taken for pipelaying operations. Where such routes cross agricultural land a year's cultivation normally results

* In May 1981 the British government announced plans to reduce the lead content in petrol by 60 per cent by 1985.

in concealment of traces; but where heather moorland is affected the litter must be carefully conserved and subsequently re-spread in order to accelerate revegetation (22).

Recreation v Conservation

It is now generally accepted that the two amenity uses of the country-side, recreation and conservation, are frequently incompatible. Con-flicts between the two are most apparent where leisure pursuits endanger sensitive sites of biological significance such as chalk grass-land, coastal dunes or mountain tops. Research on the Cairngorm Mountains, for instance, has revealed many problems in the wake of winter sports facilities having been provided. Vegetational erosion has spread along the west and southwest sides of the summit of Cairn Gorm, turves have been dug up by campers to construct sheltering walls, and rare breeding birds such as dotterels and snow buntings have been disturbed. It was suggested that some birds might well be approaching their limits of tolerance to disturbance and habitat damage at or near the ski grounds, albeit no difference in breeding successes for most species of bird could be distinguished between intensively and lightly used areas (23).

Reference has already been made to the importance of successional stage in determining the ecological capacity of a site. In addition, two further observations are of interest: certain growth forms tend to be more resilient than others; in particular, plants with an apical bud at or below ground level are less likely to be damaged by trampling than plants with buds held aloft on delicate stems — for instance, the herb layer of a deciduous woodland. Further, vulnerability may be heightened by such limiting factors as climate and nutrient poor or physically unfavourable soils. These include coastal systems (early successional stages and unstable substrate), montane habitats, now increasingly being affected by a variety of winter sports, and soils which are shallow (eg chalk grassland), nutrient poor (eg lowland heaths), or excessively wet (eg fens).

Where significant conflict between conservation and recreation is apparent, some measure of visitor control should be considered as part of an overall management plan (24). An extreme example is provided by the Galapagos Islands which are, of course, unique in ecological terms. The Islands are managed by the Charles Darwin Research Station which establishes a quota of 12,000 visitors per year to the Ecuador National Park. Allocation is largely to official tour operators; strict control is exercised over landing and access. Accommodation is generally permitted only aboard cruise ships; entry permits ensure that no more than 90 people are in any place at any time; and all visitors must be accompanied by one of the Station's highly qualified guides.

Plate 10.   An interesting example of a site where there are conflicts between intensive recreational needs and vulnerable vegetation types – at Llangorse Lake, Brecon Beacons National Park.

Resolution of Land-use Conflicts: the case of the uplands

In Britain, the 'uplands' are commonly thought of as land lying above about 250m. However, for planning purposes, the important criterion is not so much one of elevation, as a combination of bleakness, remoteness, sparseness of population and poor agricultural productivity resulting from depressed mean temperatures and increased frequency of frost. These inhospitable regions do, nevertheless, support a variety of land uses which include farming of sheep, deer and cattle, forestry (generally coniferous), wildlife and game conservation, recreation, water catchment, mineral working and defence. Whilst the last two of these uses will normally make exclusive demands upon a site, the remainder can often co-exist relatively harmoniously (25).

The major problems which resource planners must seek to resolve in the uplands are twofold (26). There is a conflict of interests between different land users, at a sub-regional scale where on the one hand, the competition for land between farming and forestry is concerned, and at a more local level in the case of recreation and conservation. Multiple use solutions have often been proposed as the panacea for this problem and, on the local scale, the simultaneous use of a tract of land for more than one purpose is indeed more feasible than in the intensive production systems of the lowlands. At the sub-regional level, however, the concept of 'integration' — which one Government report has defined as the 'co-ordination of multiple uses of land and the associated activities within a given area' (27) — would allow certain activities the exclusive use of appropriate sites. Another divergence of interests may arise between the objectives of private landowners and those of local communities: on occasion it has ben shown that landowners have been guilty of under-investment, sometimes deliberately, in order not to reduce the sporting potential of their estates.

Land use changes in these areas tend to be subject to a variety of constraints, some 'administrative', such as ownership and common land status, which may limit the availability of land for development; others are 'economic', for instance material costs and levels of government grants and subsidies. Perhaps more fundamentally, 'physical' factors place limits on the successful growth of grass and trees. Within these constraints, planners are required to use their executive and persuasive powers to the full in order to achieve that combination of activities which optimises land use, and which provides for the community the greatest level of employment, the highest income, the most balanced landscape and the greatest recreational opportunity.

Systems models

Since the relationship between land use, employment generation and community service provision is such a critical one in the marginal uplands, it is essential that planning strategies should embrace them within a truly integrated framework. A formal system model has been developed for a section of the Yorkshire Dales (28), in which inputs on land quality (classified into in-bye, rough grazing and moorland) and financial reward (based on a measure of economic rent) were analysed in order to simulate land transfers over a period of 15 years. At the end of each annual 'cycle' of the program, farms were reviewed both for possible afforestation of rough grazing land and for possible involvement in farm amalgamation, on the basis of a pre-determined rent differential.

A more informal systems approach, concentrating upon the co-ordination of policy objectives amongst land users in the North Pennines has been undertaken by the former North Riding County Council in conjunction with a number of other organisations, notably the Countryside Commission (29). A reconciliation of competing interests was achieved on the basis of multivariate statistical analysis of the responses of a representative committee of the land managers and resource development agencies; members had been asked to assign preferences and priorities to a range of alternative options for solving problems or realising opportunities within the study area. Those policy groupings which appeared to be both highly ranked and

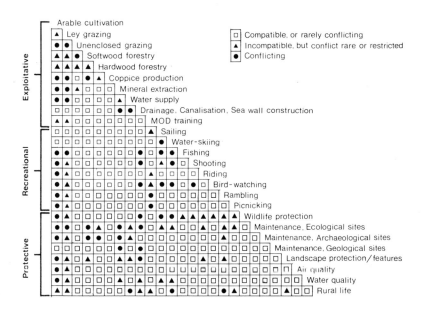

Figure 3.4 A compatibility matrix of competing interests in the countryside. (Source: Green (30).)

mutually compatible were given spatial expression, and related to the ecological potential of the environment, in the form of three alternative (non-statutory) plans.

This and other studies suggest that the scope for co-operation and agreement within the countryside and for the multi-purpose use of a great deal of rural land probably outweighs the perceived degree of conflict. Although certain uses do prove incompatible, there are still many pursuits which can co-exist. Indeed, one analysis of 26 major users of rural land shows that, of the 625 possible combinations between activities, only 53 major and 55 minor conflicts can be identified (Figure 3.4) (30).

NOTES

(1) Alice Coleman (1976) 'Is planning really necessary?', *Geographical Journal*, 142.

(2) R. Best (1976) 'The extent and growth of urban land', *Planner*, 62.

(3) See Joan Davidson and G. Wibberley (1977) *Planning and the Rural Environment*, Oxford, Pergamon, Chapter 8.

(4) K. Williams, P. Hills and D. Cope (1978) 'EIA and the Vale of Belvior Coalfield'.

(5) The Committee for Environmental Conservation has recently completed a working party report on open cast coalmining and the environment (1979) *Scar on the Landscape?*, London CoEnCo, 29 Greville St. Issues relating to deep mining are reviewed by the Royal Town Planning Institute (1979) *Coal and the Environment – a submission to the Commission on Energy and the Environment*, London RTPI, 26, Portland Place, W1.

(6) H.D. Smith, A. Hogg and D. MacGregor Hutcheson (1976) 'Scotland and offshore oil: the developing impact', *Scottish Geographical Magazine*, 92; and J. Fernie (1977) 'The development of North Sea oil and gas reserves', *Scottish Geographical Magazine*, 93.

(7) How planners and conservationists began to tackle the potentials and problems of the North Sea discoveries in the early years is outlined in the theme feature 'Oil: questionable benefits for Scotland' in *Built Environment*, May 1973, Volume 2, No 5.

(8) Scottish Development Department (1974) *North Sea Oil and Gas: coastal planning Guides*, Edinburgh, HMSO. See also a review of environmental protection measures in P.H. Selman (1977) 'Planning for environmental quality', *Scottish Geographical Magazine*, 93.

(9) Judge Parker (1978) *The Windscale Inquiry* (2 vols), London, HMSO.

(10) P.H. Selman (1979) 'Pumped storage electricity — the latest threat to the uplands?', *Planner*, 65.

(11) P. Chapman (1975) *Fuel's Paradise: energy options for Britain*, Harmondsworth, Penguin.

(12) See J.R. Blunden (1975) *The Mineral Resources of Britain: a study in exploitation and planning*, London, Hutchinson.

(13) H. Godwin and S.M. Walters (1967) 'The scientific importance of Upper Teesdale', Proceedings of the Botanical Society of the British Isles, 6.

(14) A useful summary of these and other proposals is given in: Open University, Earth's Physical Resources Course Team (1974) *Water Resources*, Milton Keynes, Open University Press.

(15) F. R. Harden-Jones (1968) *Fish Migration*, London, Edward Arnold.
(16) J. R. Jefferson (Chairman) (1976) *Route Location with Regard to Environmental Issues* — report of a working party, London, Department of the Environment. For general discussion see: D. Lovejoy (1973) 'Communication systems in the landscape' in D. Lovejoy (Ed.) *Land Use and Landscape Planning*, Aylesbury, Leonard Hill.
(17) Commission on the Third London Airport (Chairman Lord Roskill) (1970) Papers and Proceedings, Volume 7, London, HMSO.
(18) J. Edington and Ann Edington (1977) *Ecology and Environmental Planning*, London, Chapman and Hall.
(19) M. W. A. Cassidy (1976) 'Lessons to be learnt from case studies in London' in T. O'Riordan and R. Hey (Eds.) *Environmental Impact Assessment*, Farnborough, Saxon House.
(20) J. Mathews (1978) 'The lead battle enters the courts', *New Scientist*, Vol 79.
(21) D. J. Jeffries and M. C. French (1971) 'Lead contamination in small mammals trapped on roadside verges and field sites', *Environmental Pollution*, 3; J. H. Williamson (1973) 'Lead in roadside vegetation': paper presented to *Motorways and the Biologist Symposium*, North East London Polytechnic, 25th October 1973.
(22) D. A. Gillham and P. D. Putwain (1977) 'Restoring moorland disturbed by pipeline installation', *Landscape Design*, 119.
(23) Joy Tivy (1972) *The Concept and Determination of Carrying Capacity of Recreational Land in the USA*, Countryside Commission for Scotland, Occasional Paper No 3.
(24) A detailed account of the management plan for a local Nature Reserve is to be found in M. B. Usher (1973) *Biological Management and Conservation*, London, Chapman and Hall.
(25) See, for example, P. H. Selman (1978) 'Alternative approaches to the multiple use of uplands', *Town Planning Review*, 49, and Joan Davidson and G. Wibberley *op. cit.* (3), Chapters 9 and 10.
(26) W. Ellison (Chairman) (1966) *Forestry, Agriculture and the Multiple Use of Land*, London, Department of Education and Science.
(27) A. O. Dye (1973) 'Upland sub-regional planning using a simulation model', *Journal of Environmental Management*, 1.
(28) D. Statham (1972) 'Natural resources in the uplands — capability analysis in the North Yorkshire Moors', *Journal of the Royal Town Planning Institute*, 58.
(29) North Riding County Council (1975) *North Riding Pennines Study: Study Report*, Northallerton, North Riding County Council.
(30) B. Green (1977) 'Countryside planning: compromise or conflict', *Planner*, 63.

4 Resource Planning — containment of waste and pollution

It was noted in the earlier discussion of ecological energetics (page 52) that energy transfers between trophic levels were rather inefficient. This net energy loss is due partly to the creation of waste products. In extreme cases, wastes may become so concentrated locally that they interfere with the health of animal populations and thus behave as a density dependent curb on their size.

Nowadays, traces of man-induced toxins can be detected everywhere in the world, although this is rather a testimony to the sensitivity of modern analytical techniques than to man's depredations. It has been widely suggested that man's continued contamination of the earth by harmful pollutants could, in due course, reduce the capacity of the environment to sustain him. Whether this is taken as proof of the profligacy and myopia of the human species or whether pollution is accepted as a natural by-product of man as a creative animal, the planner's response should be a determined and constructive approach towards the safe disposal of pollutants and to the recycling of wastes.

Industrial Solid Wastes

Of all aspects of pollution treatment, the reclamation of land despoiled by industrial by-products has probably been the one which has concerned planners to the greatest degree. The most extensive damage over the country as a whole has been caused by the deposition of spoil from the coal mining industry, most particularly in north England, the north Midlands, south Wales and parts of central Scotland. Other parts of the country, however, have their own characteristic forms of land pollution. Areas which were the centres of the iron industry have inherited huge tips of slag in the wake of the closure of the old blast furnaces; large tracts of land have been blighted with toxic wastes from the chemical industries, notably along the Mersey-Irwell basin; in Cornwall and Devon the extraction of china clay has left behind nine

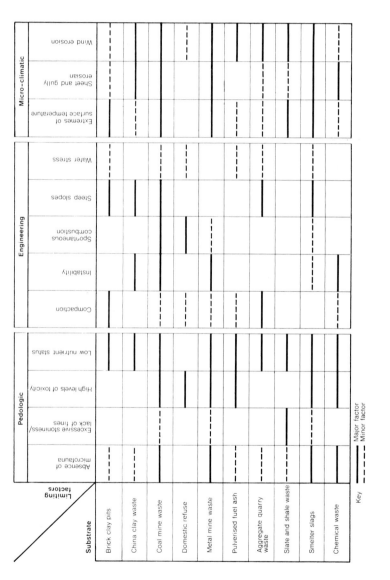

Figure 4.1 Factors limiting successful plant growth on various types of derelict land. (Adapted from Goodman (2).)

parts of waste to every one of clay extracted; whilst the quarrying of slate has despoiled tracts of magnificent countryside with its huge and stark tips. Today, new processes, such as the production of ash from power stations which burn pulverised coal, contribute to this legacy of despoliation.

In combatting land pollution, the planner has the choice of three courses of action. (a) Conditions of planning consent may be attached to a new development, requiring the operator to minimise future dereliction. In such cases, matters of safety are likely to take precedence over those of amenity, which is why the modern tips of waste from coal extraction tend to be featureless and flat-topped in outline, spreading over much larger areas of land than in earlier decades. (b) Conventional methods of landscaping may be employed, in which old tips are contoured or flattened and the most noxious wastes buried beneath the less harmful deposits or topsoil. (c) Ecological principles may be applied in order to achieve the revegetation of solid wastes, sometimes with lower costs and greater long-term success than conventional methods, although their after-use will normally be restricted to amenity purposes (1).

Site restoration

Typically, a wide variety of physical and ecological factors will combine on each site to render it at least partially unresponsive to revegetation (Figure 4.1) (2). Depending on the severity of these properties, the ecologist may adopt one of two approaches, either accepting the unfavourable site conditions as they are and experimenting with pioneer plant species which seem likely to have a high tolerance, or attempting to incorporate a soil amendment into the substrate in order to ameliorate it. Hence, before recommending a course of treatment, an ecologist will normally require to know a variety of details about the site, including the adequacy of its current nutrient status and its ability to retain nutrient amendments, the variability of pH values over the site, the abundance of trace elements (such as copper, zinc and lead) which may be present in toxic concentrations and the presence of dissolved salts which may interfere with the osmotic balance of plants. In addition, attention must be paid to the shape, slope and aspect of a site, features which will markedly affect its rate of natural colonisation.

Where it appears that a succession can be initiated without the need for costly reclamatory techniques, species may be selected according to their ability to tolerate specific adverse conditions. Certain grasses, such as common bent (*Agrostis tenuis*) and the fescues *(Festuca* spp.), which have low nitrogen and phosphate requirements, can survive on sites deficient in these nutrients. Others may be tolerant of extreme pH

Plate 11.   Land requirement for waste disposal is steadily increasing – a problem compounded by fly-tipping.

values, for example wavy hair-grass (*Deschampsia flexuosa*) which has been found growing on acid colliery spoils of pH3·2, and red fescue, which will grow on pulverised fuel ash and blast furnace slags, where alkalinity may exceed pH8·5. Some plants may withstand dissolved salts, including, not surprisingly, many marine species, and these will grow readily on saline industrial wastes. Physical and moisture conditions, such as steep slopes, waterlogging, rocky and stony substrates and ground compaction, rather than chemical abnormalities, may frequently be the limiting factors, and here alders and willows are often successful, although semi-aquatic species may prove necessary in extreme conditions (3).

Certain wastes, notably slate, being of a very hard, durable, waterproof character, weather into a soil only very slowly. Moreover, the lack of organic matter and the inability of the slate to retain water will, in the absence of adequate quantities of fine shale, create drought conditions at the surface for much of the year, which will further reduce the likelihood of seedling establishment. Reclamation techniques for slate most often require the planting of trees in relatively stable pockets filled with a nutrient-rich medium which will absorb and retain water (4).

Frequently, substrates will be so unstable that conventional seeding methods would be impracticable or impossible. In these circumstances, it is becoming increasingly popular to establish grass and other vegetation directly, by using high pressure equipment to spray onto the site a liquid emulsion containing mulch, seed, fertilisers, micro-organisms, trace elements and soil bacteria, together with a protective bulking medium. This process is most useful where steep, inaccessible and sometime non-topsoiled slopes — of shale, pulverised fuel ash, household refuse and motorway embankments — must be treated.

Waste disposal

A continuing problem faced by local authorities, and an increasingly difficult responsibility to comply with, is the disposal of solid wastes. In addition to the 18m. tonnes of domestic and commercial refuse which must be disposed of annually by local authorities, a further 5m. tonnes of trade waste and 15-20m. tonnes of industrial waste are produced, and whilst these are generally disposed of by private contractors, local authorities are responsible for coordinating such operations via planning controls and licensing procedures. With progressively fewer holes in the ground available for infilling, much dumping must take place in sites of ecological interest such as abandoned quarries or valley bogs. Attendant on this is the risk of the lateral migration of toxic elements from such landfill sites into groundwater supply strata.

A recent canvass of 22 disposal authorities in England and Wales showed that 13 were having difficulty in finding sufficient landfill sites to dispose of their own wastes (5). This suggests a need to find alternative, and perhaps more constructive, methods of dealing with our refuse if this increasing source of land pollution is to be contained in the future. Incineration, possibly with the recovery of heat for district heating systems or electricity generation, and reclamation have both been attempted on a small scale; but their success has been restricted by the nature of modern wastes, notably plastics, which are difficult to burn or recycle. As an alternative, Merseyside County Council is experimenting with tipping of suitable materials as a means of improving the drainage of potentially good agricultural land after carefully stripping and storing the topsoil.

Radioactivity

Radiation is not something artificial or alien to life. Indeed we are all subjected to background levels of radiation arising from cosmic rays, rocks and human and animal bodies. However, nuclear radiation may become a serious form of pollution in either of two ways: in the event of man's contributions to background levels becoming far greater which is improbable, or, more likely, the significant increase of levels at a specific local source. Moderately low intensities may cause no detectable effects in a single dose, but if there are repeated exposures, harmful effects may result and it is therefore difficult if not impossible, to establish a level below which radioactivity can be assumed completely harmless (6). So far, the control of radioactive pollutants has centred upon the safe storage and carefully regulated use of materials. However, it has been suggested that hazards to society as a whole might arise from widespread generation of nuclear electricity. Increasingly, aspects of a future nuclear programme, notably site safety and waste disposal, will give rise to material planning issues.

Gaseous and Particulate Emissions

The burning of fossil fuels, particularly coal and oil, is the greatest single cause of air pollution. Coal produces many particles of varying sizes, the larger making up dust and the smaller smoke. Dust is usually deposited near to its source and may cause little deleterious effect beyond the blackening of buildings; but smoke particles remain longer in the air and may be breathed into the lungs becoming particularly dangerous when they form a constituent of 'smog'. Sulphur, the other main pollutant from fossil fuels, is largely emitted as sulphur dioxide gas (SO_2), which tends to combine with the excess of ammonia, also present, to produce ammonium sulphate which, although far less toxic, is slightly corrosive to masonry.

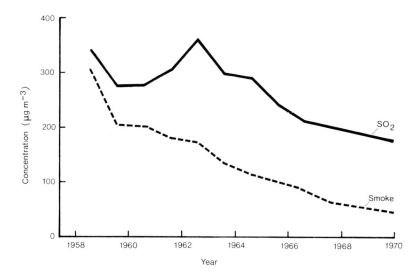

Figure 4.2   Concentrations of two major components of atmospheric pollution, sulphor dioxide and smoke, in London air. (Adapted from Committee of the Royal College of Physicians of london on Smoking and Atmospheric pollution, 1970 Summary and Report on Air Pollution and its effect on Health, Pitman, London.)

For some years now, levels of SO_2, particulate matter and certain ions in the atmosphere have been monitored by the Warren Spring Labortory. Although the distribution of these stations is very uneven, it appears that smoke concentrations have declined markedly in urban areas in the wake of increasingly effective legislation; however, the reduction of SO_2 levels has been less effective (Figure 4.2). In rural areas, pollution seems to have increased somewhat as a result of high altitude emissions from industrial and generating centres.

The concentration of airborne gases and particulate matter may become locally severe, especially where adverse microclimatic or topographic conditions prevail. In particular, temperature inversions, in which warm air overlies cold so that buoyancy forces inhibit vertical movement (see Ch. 1), may entrap pollutants in a dense fog close to the ground. Clark *et al.* (7) have summarised some of the more unexpected local effects in pollution patterns caused by topography: a wooded valley close to the centre of Newcastle had much lower levels of SO_2 than the surrounding area, but particulate matter levels were little reduced; around the Port Talbot steel works, an unusual valley system channelled polluted air from its source in the valley bottom, with resultant pollutant damage to plants where it flowed over the crests of the enclosing hills. Similar instances of fumes being trapped by restrictive topography have also been observed around aluminium smelters at the entrance to Norwegian fjords. Careful analysis of local microclimates could greatly assist in the choice of sites for industry; locations having optimum conditions for the harmless dispersal of airborne emissions would plainly qualify as the most suitable for the siting of heavy industry.

The normal method of reducing local pollutant concentrations is for planners or environmental health officers to stipulate the height of relevant chimney stacks, in order that gases and particulates are released above the level at which inversions commonly occur. However, when inversions form aloft the stack, or when very strong lapse conditions prevail, dispersion may still be unsatisfactory. In addition, pollution is not prevented, but the harmful agents merely scattered more widely: low concentrations of SO_2 effected in such a manner may combine with water droplets to produce an 'acid rain' which may fall over large areas, often many miles from the source. Thus, the acidification of soils and streams in Scandinavia, possibly affecting the long term productivity of the northern coniferous forests and the salmonid lakes, has been attributed to high level emissions in the industrial regions of Western Europe.

Plants and trees are generally more susceptible to atmospheric pollution than are animals, and may register acute responses to high levels of pollution over short periods of time (eg $750\,\mu g\ SO_2$ per cubic metre of air for a 6 to 24 hour period), chronic responses to lower levels over longer periods of time (eg $150\,\mu g\ SO_2\ m^{-3}$ for several days), or subchronic or cryptic responses to very low concentrations over long periods of time (eg $50\,\mu g\ SO_2$ for several weeks or months). Nowadays, the tolerances of different species to varying concentrations of airborne pollutants are quite well-established, enabling landscaping schemes to be carried out with a greatly increased probability of success (8). Where reliable data on SO_2 concentrations are not available, an adequate proxy measure may be obtained by recording the presence or absence of certain sensitive indicator species, notably selected lichens and fungi. Particulate matter, for instance from quarrying, may also have an adverse effect on plants, especially if thick layers of dust accummulate on leaves, restricting photosynthesis.

Another pollutant of soil and vegetation is fluorine, which tends to be deposited close to aluminium smelters and steel works which use it as flux and to brickworks in which clays containing fluorine compounds are used. Fluorine in the soil may be absorbed by plant roots, or leaves may absorb a surface deposit of particulate matter containing it; the combined effect of these two mechanisms will result in plants concentrating significantly higher levels than prevail in the ambient air. Fluorine may be transmitted through ecological grazing chains and many cases of fluorosis among dairy cows and other commercial herbivores have been documented.

Plate 12. A depressing scene of land, water and air pollution and visual intrusion. Hopefully, modern planning measures will succeed in creating industrial developments in which these are kept to a minimum.

Organic Compounds

The most noted type of pollution by persistent organic compounds has probably been that of the sea and coastline by the accidental or deliberate spillage of oil (9). The sheer scale of the international transportation of petroleum and of its exploitation at sea, combined with the paucity of effective legislation regarding liability in international waters, makes the occurrence of pollution almost inevitable. The major types of accident have been occasioned by incidents concerning the 'blow-out' of offshore oil wells and the collision of oil-carrying supertankers. The worst offenders in this respect have been oils with a heavy tar residue, such as Middle Eastern crudes; oils so far recovered from the North Sea should not present such risks of coastline fouling, although they do contain more toxins than the heavier crudes.

The effects of oil pollution have been particularly detrimental to wildlife. The feathers of sea birds get clogged, leaving them poorly insulated and unable to fly; also they ingest the poisonous oil whilst attempting to clean their plumage. However, the natural processes of bacterial action normally ensure that oil is broken down fairly rapidly unless its concentration is too high, and in practice the most serious damage to wildlife has often been caused by the detergents used to clean up the oil itself. Less dangerous dispersants and detergents are becoming available and it is essential that, wherever developments which might lead to oil pollution are being contemplated, planners should ascertain from the operator which compounds he intends to use in the event of an emergency.

In addition to the more spectacular accidents, intentional spillages, small but incremental, occur when tankers deliberately discharge waste oil and also when they are washed out prior to loading. The frequency of this practice has lessened in recent years with the general adoption of the 'load-on-top' system, but nevertheless spillages do still lead to a continuous state of oil-fouled water in ports. Fischer and von Winterfeldt (10) have discussed the regulations for controlling operational ('chronic') discharges in the North Sea where routine emissions of washing and ballast water from ships, and of displacement water and production water associated with storage tanks, and oil-water separation processes from production platforms, occur. The setting of standards is apparently achieved with more administrative impartiality by the Norwegian government than by the British, although the levels set by both are similar (approximately 30ppm of petroleum in discharged water). In both cases, due to an insufficient body of research on the biological status of production areas, the standards must be related to current pollution control technology rather than to ecological tolerances, and so there is little evidence of their adequacy in relation to the long-term effects of chronic discharges.

Oil pollution from landward sources may also become a problem, especially in the United States and the European mainland where oil is transported by pipeline to refineries located adjacent to inland industrial centres. Pipelines are regularly subjected to a 'line balance' to determine any discrepancy between input and output quantities, and so leakages tend to be quickly located and rectified, although far greater risks arise when pipelines must cross waterways or shipping channels. In addition, institutions using fuel oils, such as garages, factories and hospitals, may be responsible for significant localised pollution if the handling and disposal of oil is inexpert.

Pesticides

Perhaps the most serious form of land-based pollution by persistent organic compounds has been that caused by pesticides. A pesticide is normally used against a particular organism and ideally it should poison that organism only and otherwise be harmless; this degree of selectivity is often impossible to achieve, however, and the testing of products for unwanted side-effects is often inadequate.

Many pesticides are very persistent in their behaviour and continue to have undesirable and unintended effects on wildlife for very long periods, most notably the organo-chlorine insecticides, including DDT, Lindane and members of the Cyclodiene group (eg Dieldrin and Aldrin). Most of these compounds are not strictly controlled in their application, and, although voluntary restrictions on the usage and availablity of the more lethal and less rapidly degraded amongst them exist, these have been known to be breached (11). Whilst local authorities can exert no direct control over the application of pesticides, their patterns of use and their side-effects remain of interest to planners as one major aspect of the contemporary management of rural land.

Sewage

Pesticides may be regarded as 'biologically resistant' wastes since they are poisonous to life. Conversely, certain organic wastes are 'biologically available' and may be broken down by micro-organisms which utilise oxygen in the process. The most widespread example of pollution caused in this manner has been the deoxygenation of inland and estuarine waters by discharges of inadequately treated sewage — and this is clearly an area which is very much within the control of the local authority. The quantity of organic matter present is determined by the biochemical oxygen demand (BOD) of the water, which is measured by a test showing the ability of the sample to absorb oxygen; thus the higher the level of absorption, the greater the degree of pollution. Where BOD is particularly high, a reducing, anaerobic environment will occur in which only a few organisms are able to survive, such as bacteria, sewage fungus and resistant animals like *Tubifex* worms (12).

Today, most sewage in Britain is treated before discharge to inland waters, although it is regrettable that standards of 'secondary treatment' are not uniformly high. Of rather greater concern, in certain parts of the country, is the problem of farm animal wastes, for, with an increasing proportion of cattle, pigs and poultry now being largely confined to buildings, their wastes become a form of pollution rather than a useful fertiliser. Facilities for the treatment of such wastes are frequently minimal or non-existent, and this raises issues of whether agricultural operations should not be at least partially subject to planning control.

In addition to the manner in which short-life wastes can raise BOD levels, two further side-effects should be noted. One is that when sewage is filtered, nutrient salts may be passed out, causing greatly accelerated eutrophication of slowly-moving or standing water; and second, effluents may contain suspended solids capable of smothering aquatic life and restricting the penetration of sunlight. Commonly, water authorities seek to ensure that effluents discharged into confined waters contain no more than 20ppm suspended solids and that their BOD does not exceed 30 — the so-called '20/30' standard. However, these were limits set many decades ago, and it could be argued that they fail to meet modern expectations of amenity.

Thermal Pollution

The heating of rivers or inshore waters may, on occasion, reduce the amount of aquatic life sufficiently to qualify as a form of pollution. Whereas warm-blooded animals have developed homeostatic mechanisms which maintain a nearly constant body heat, aquatic life, being *poikilothermic*, is highly sensitive to changes in ambient temperatures. In addition, as water is heated, so the solubility of oxygen declines and yet the rate of metabolic activity of the aquatic organisms is raised, thus increasing oxygen intake requirements.

There have been reports from abroad of river water being heated to 50°C and above, so that only a few thermophyllic bacteria are able to survive; the problem has so far been far less acute in Britain, although some thermal pollution does occur, particularly in association with electricity station cooling towers. At the Drakelow power station on the River Trent, the Central Electricity Generating Board have found that the average summer temperatures of water have been raised from 18·2°C above the inlet to 27·4°C below the outlet, and the winter temperatures from 6·6°C to 20·4°C. At this site, it was recorded that the *Tubifex* population was greatly increased and that there was a corresponding increase in the length of their breeding season. On other occasions, the replacement of game fish by coarse species has been observed.

Where such phenomena occur at the coast, proposals have been made to establish shellfish farms: however, as many of these stations are for the production of nuclear power, and since molluscs appear to be highly efficient concentrators of radio-active effluent, the desirability of any shellfish so reared might be reduced by fears of possible contamination.

NOTES

(1) See, for instance, R. Gemmell (1977) *Colonisation of Industrial Wasteland*, London, Edward Arnold.
(2) Reviewed in G. T. Goodman (1974) 'Ecological aspects of the reclamation of derelict land' in A. Warren and F. B. Goldsmith (Eds.) *Conservation in Practice*, London, Wiley.
(3) J. A. Richardson (1976) 'Pit heap into pasture: natural and artificial revegetation of coal mine wastes' in J. Lenihan and W. W. Fletcher (Eds.) *Reclamation*, Glasgow, Blackie.
(4) J. C. Sheldon and A. D. Bradshaw (1976) 'The reclamation of slate waste tips by tree planting', *Landscape Design*, 113.
(5) B. Dodd (1979) 'Waste disposal: a mounting costly headache for local authorities', *The Planner*, 65.
(6) K. Mellanby (1972) *The Biology of Pollution*, London, Edward Arnold.
(7) B. D. Clark, K. Chapman, R. Bisset and P. Wathern (1976) *Assessment of Major Industrial Applications: a manual*, Department of the Environment Research Report 13 (see Appendix E), London, HMSO.
(8) J. F. Benson and R. J. Bevan (1977) 'The use of biological indicators of air pollution as an aid to species selection in landscape design', *Landscape Design*, 119.
(9) For a fuller discussion of this topic see A. Nelson Smith (1972) *Oil Pollution and Marine Ecology*, London, Paul Elek Ltd.
(10) D. W. Fisher and D. von Winterfeldt (1978) 'Setting standards for chronic oil discharges in the North Sea', *Journal of Environmental Management*, 7.
(11) See, for instance, K. Mellanby (1970) *Pesticides and Pollution*. Revised edition, London, Collins.
(12) H. B. N. Hynes (1960) *The Biology of Polluted Waters*, University of Liverpool Press.

5 Land Use — some techniques for environmental planning

Land use planners are concerned with securing the right development in the right place at the right time; their primary requirement in this task is for a range of survey and analytical techniques, to ensure that land use decisions are made in the light of the fullest possible information. Unfortunately, most of the techniques devised so far have been concerned with metropolitan systems, and rural resources have at best been considered as negative constraints rather than as positive elements in their own right. If we are to secure the rational use of ecological resources, we must develop, on the one hand, methods of survey and evaluation which define the intrinsic resource potential (land capability) of a unit of land in relation to certain pre-determined criteria (Table 5.1) (1) and, on the other hand, optimisation techniques which seek to allocate the most appropriate uses to units of land on the basis of this potential.

Qualitative dimension	Features to be assessed
1. Agriculture	Fitness of land resource for agricultural production
2. Forestry	Fitness of land resource for forestry production
3. Water	Capacity of land to supply, store and drain water to acceptable standard of purity
4. Landscape	Aesthetic quality of the countryside
5. Informal recreation	Capacity of countryside to support informal recreation
6. Nature conservation	Capacity of countryside to sustain wildlife and other features of nature conservation interest

Table 5.1   Factors in land classification. (Adapted from Gane (1).)

Environmental specialists have frequently underestimated the importance of evaluation and optimisation in the planning process, and they have proved reluctant to provide planners with straightforward and readily comprehensible resource assessments. This reluctance may stem from an awareness of the pitfalls inherent in reducing to an ordinal scale those complex factors which determine relative value, and planners must certainly be cautious in the interpretation of simple indices of resource quality. Similarly, very considerable problems arise when trying to compare assessments for different resources: for instance, in what sense may top quality farmland be considered comparable with outstanding landscape? Despite the intractability of such problems, however, the further development and utilisation of resource planning techniques is essential, for it is generally to the planner that the often invidious task of arbitrating between different land users falls (2).

Techniques for Survey and Evaluation

Agriculture

Perhaps the single most important aspect of planning intervention in the countryside lies in ensuring that good agricultural land does not pass into some less productive use. In the lowlands this normally entails safeguarding farmland from urban encroachment, whilst in the uplands, where planning intervention tends to be less direct, the main priority lies in encouraging the co-ordination of land transfers between marginal agriculture and forestry. For both these purposes, an evaluation of relative agricultural productivity is necessary.

Three map series dealing with land quality are currently available. The most comprehensive in its coverage is the *Agricultural Land Classification Series*, which is prepared by the Ministry of Agriculture, and covers the whole of England and Wales (a comparable survey of the more populous areas of Scotland has been carried out by the Department of Agriculture and Fisheries for Scotland, although it remains unpublished and is available for inspection only at the Department's regional offices). In addition, there are two *Land Use Capability Series* published by the Ordnance Survey on the basis of work carried out by the Soil Surveys of England and Wales, and of Scotland, respectively.

These classifications differ in detail but the principles upon which they are based are similar (3). Essentially, the assumption is made that, since farming operates in a context of changing prices and costs, the flexibility of a unit of land to produce economic yields of a wide range of crops will be the principal determinant of land quality. Only soils of the very highest quality can support the most demanding crops, and most soils will be restricted in their usage by a variety of 'limiting factors' which serve to restrict the possible crop combinations. Land excluded from the highest grades should not, therefore, automatically be deemed unworthy of protection, for it may often be capable of producing high yields from other less exacting uses, such as dairying, and there has indeed been strong criticism of current appraisal techniques for their undue emphasis on agronomic rather than economic crtieria (4).

The two Land Capability series provide rather technical assessments of likely crop performance, based on the detailed investigations of the Soil Survey. Land is assigned to one of seven classes, classes 1-4 including land suitable for crops, classes 5 and 6 comprising land more suited to grazing and forestry, whilst class 7 land is normally of use only for recreation and game conservation (Table 5.2) (3). Once land has been placed in a primary class, further notation may be added in the form of sub-scripts indicating the presence of particular limiting factors, namely soil deficiencies (s), wetness and drainage problems (w), adverse climate (c), liability to erosion (e) and excessive gradient (g). These surveys are essentially intended to assist the farmer and his technical advisers and may not be immediately suitable as an information base for planners. However, a specially commissioned capability classification of the Island of Mull has been utilised by the Highlands and Islands Development Board as a basis for hill land improvement together with an integrated programme of afforestation (5).

The land classification scheme of the Ministry of Agriculture in response to the continued loss of farming land to urban and other development, has been prepared with the needs of planners more specifically in mind. The scale at which it is mapped (1:63360 or 1:50000) at times raises problems of detailed interpretation, although this is offset by its comprehensive coverage and the (supposed) consistency with which land is classified nationally (6).

The classification was specially devised so that it should be capable of clear definition and be readily comprehensible to users not necessarily expert in agricultural matters: thus, there has been an attempt to make it as uncomplicated as possible by separating land into five objectively defined grades. It is estimated that Grades 1 and 2, those with the greatest flexibility of cropping, account for only about one-sixth of the agricultural land of England and Wales and are proportionately even scarcer in Scotland. These areas of high-yield, low-cost farming are invariably found on well-drained or gently-sloping lowlands and as such they are also likely to prove the most attractive to the building developer. Grade 3 land accounts for almost half the farmland of England and Wales, embracing as it does land which is just marginally poorer than that with 'minor limitations' (Grade 2) and land just slightly superior to that having 'severe

Land Use Capability Class	Degree of Limitation	Principal Agricultural	Principal Enterprises	Levels of Yield
1	Very minor	All usual British crops	Arable, horticultural	High
2	Minor	All usual British crops, but increased risk of failure for horticultural produce	Arable, horticultural	High yields with good management
3	Moderate	Cereals, grass, forage crops, potatoes	Arable, dairy	High yields from selected crops under good management
4	Moderately severe	Grass dom-dominant, restricted cereals	Dairy, livestock (rearing and fattening), forestry, limited arable	High yields of a very restricted range of crops possible under good management
5	Severe	Improved grass	Dairy, livestock rearing, forestry, recreation	High yield of grass products possible, but risk of failure high
6	Very severe	No cropping, improvement very difficult	Extensive livestock rearing, some forestry, recreation	Moderate yields; care needed in grazing management to prevent sward deterioration
7	Extremely severe	No cropping	Some grazing by hardier stock, recreation	Low yields, very short season

Table 5.2   Characteristics of land classes in the Land Use Capability Classification. (Adapted from Bibby & Mackney (3).)

limitations' (Grade 4); a sub-division of this grade is now being introduced.

There are doubts as to how seriously planners have taken the long-term significance of good farmland despite instructions from successive environment ministers to treat it as a matter of priority. One study of development control decisions in Devon seems to suggest that the ratio of permissions to refusals is unrelated to the agricultural quality of the sites affected — apparently rural land is being treated in blanket fashion rather than on a scale of priorities (7). If this is generally the case, and there is evidence to suggest that it may be, then it indicates an approach which cannot be afforded in a situation of resource scarcity.

Forestry

Whereas the agricultural potential of a unit of land may be assessed purely on the basis of its physical limitations, the suitability of land for forestry must also pay regard to constraints of an economic nature. Thus, whilst exposure and poor soils impose a limit of absolute unprofitability upon timber-growing operations, higher quality land sets its own margins of relative unprofitability when compared with agriculture. Capability classifications for forestry therefore tend to concentrate in detail on land lying between these two confines and which tends to be grouped together in a rather undifferentiated manner for agricultural purposes. In practice, expansion of the national forest is restricted to areas defined in the Land Use Survey as 'moorland, heath and rough grazing', although the private forestry sector is not so constrained in its uptake of land.

Two alternative approaches to the silvicultural assessment of land are commonly used, one being based upon the yield characteristics of the tree crop ('growth classification'), and the other, which is of rather greater interest to land use planners, emphasising site characteristics, such as climate, soil or vegetation ('site classification') (8). An ecological approach to site evaluation has been devised for use in the Canada Land Inventory (9), and its application in this country has frequently been advocated. It must be borne in mind, however, that Canada has extensive reserves of virgin forest, for which the principle of altering limitations or substituting species has not been generally adopted, and which provide a reservoir of information on species performance in relation to site. Such conditions do not prevail in Britain where the evaluation of site potential must largely be made in the absence of empirical data; this task is further complicated by our limited knowledge of the 'autecology' of the various commercial species which are commonly planted. Moreover, the likely effects of drainage, cultivation and fertilisation, and indeed of the tree crop itself, prove difficult to predict, as they ameliorate the site over varying periods of time (10).

In Britain, there is no national evaluative programme for the assessment of forestry potential, although the Forestry Commission conducts its own soil survey and site assessment for selected areas. It has been found that, whereas small-scale factors such as slope and stoniness are of primary significance to the farmer, it tends to be the broader climatic and topographic factors which critically affect tree crops. Thus, initially, the topographic and climatic setting of the site is analysed in terms of its precipitation surplus (excess of precipitation over evapo-transpiration) and relative topographic exposure. This latter factor is measured by the 'topex' value, which represents, for a particular locality, the sum of the angles of inclination to the horizon at the eight major points of the compass; consequently, the lower the topex score, the higher the exposure of the site relative to nearby locations. Then, the nature of those features of parent materials and soils most relevant to silviculture, notably soil depth and rootable depth, must be recorded. Finally, an assessment of 'windthrow' hazard must be made, in order to establish the likelihood of trees being uprooted by gales at various stages of their stand cycle (11).

The Forestry Commission's surveys are, however, restricted in their availability to areas under current consideration for planting; planners concerned with the future development of the uplands must often resort to alternative sources of information. One such approach has been used as part of a broader analysis of land capability in the North Yorkshire Moors (12). It was assumed that afforestation would not be viable on land in Grades 1-3 of the Agricultural Land Classification, that existing investment in agriculture (for instance, buildings, stone walls, etc.) would lower the economic potential for forestry even where the ecological potential was high, and that the existence of common land would prove at least a short-term constraint. By applying these principles, and consulting the local officers of the Foresty Commission, a preliminary assessment of silvicultural capability was obtained (Table 5.3) (12). Ideally, some consideration should also be given to non-physical factors such as land prices, access, ownership and amenity but, in the interests of simplicity, these are often excluded or re-introduced in a subsequent analysis.

Wildlife

The evaluation of land for agriculture or forestry relates to a fairly simple criterion, namely, the level of yield which may be expected from a given area under specified management practices. However, the criteria which are relevant to wildlife quality are less obvious, for, not only does a wildlife survey involve the mapping of interconnected and dynamic ecosystems on a discrete and static basis, but its assessment requires the assignment of objective values to elements which are at

Grade	Soil Characteristics	Economic con- straints to Forestry Development	Other Constraints
1	Variable but generally poor soils	Low	
2	Poor to very poor soils	Low to moderate	
2a (common land)			Legal and administrative
3	Poor soils	Moderate to severe	
4	Moderate to good	Severe con- straints, except for sporting and amenity purposes	
5			Severe ecological constraints

Table 5.3 A Planning classification of forestry potential. (Adapted from Statham (12).)

least partially unquantifiable in their worth.

However, where ecological evaluations are available, they may be used to influence policies at both a structure plan level, where pressures may be channelled away from sensitive areas, and at a local level, where development may be controlled in order to protect sites of conservation value. For detailed interpretive work, therefore, a base map showing the precise distribution and characteristics of vegetation types should be prepared, whilst from this should be abstracted a generalised map of broad ecological zones expressed in terms of their relative value.

In the past, assessments have tended to involve the compilation of detailed species lists for selected sites but it has more recently been demonstrated that the mapping of habitat types, which takes only a fraction of the time, will produce estimates of quality sufficiently accurate for most planning purposes. Once the botanical features of a region have been recorded, an attempt is made to assign scores to sites which reflect their relative conservation values, these being determined on the basis of a variety of criteria, most notably rarity and diversity. Other factors, however, such as history of documentation, vulnerability, representativeness of a habitat type and spatial continuity may also be important (13).

One of the first applications of this method in Britain was undertaken by the Nature Conservancy (now the Nature Conservancy

Council) on behalf of Hampshire County Council (14). Once the surveyors had prepared a generalised botanical map of the county, a zonation of this into homogeneous units was derived according to the assumption that a tract of land would be broadly characterised by the dominance of one of three categories of vegetation related directly to land use: agricultural land, plantation woodland and 'unsown' or semi-natural vegetation.

The scoring of the zones then proceeded according to three principles: that, in view of their very limited distribution in lowland Britain and of the intensity of development and reclamation pressures which they face, semi-natural habitats could automatically be accorded a high value; that within such a highly cultivated area, plantation woodlands would often form relatively important wildlife reservoirs; and that the conservation interest of farmland would vary in inverse proportion to its intensity of usage. On this basis unsown vegetation was placed in categories I or II, and plantation woodland into categories II or III. Agricultural land was deemed to have a relative value between II and V, depending on the relative abundance or absence of selected landscape features such as hedgerows, ponds and reservoirs.

There is a frequent possibility of site mapping in rather greater detail, particularly when experienced volunteers can be found to assist at the surveying stage, which enables more precise ecological advice to be given at short notice. The Nature Conservancy Council in Scotland, for instance, is conducting a countryside survey with the assistance of volunteers from the Scottish Wildlife Trust. Each vegetation type is recorded by a colour and letter code which indicate the primary site class and its secondary characteristics, respectively. Once the habitat map has been completed, it then becomes possible to score each site on a trial-and-error basis, although if time permits, the percentage regional cover of each habitat can be measured and a value, or conservation priority, assigned according to its relative scarcity.

Of considerable interest to planners is the recently completed *Nature Conservation Review*, which now provides a comprehensive national record of areas of biological significance (15). Described in it are the features of more than 700 major sites, together with an inventory of many more minor ones, classified into eight main ecosystem groups (coasts, woodlands, lowland grasslands, heaths and scrub, open water, peatlands, uplands, and artificial). The habitats are scored on a six-point qualitative scale, on which Grades 1 and 2 denote 'key sites' of national and sometimes international importance (and often worthy of purchase as reserves), whilst Grades 3 and 4, of which several thousand are identified, are more appropriate for designation as Sites of Special Scientific Interest; Grades 5 and 6 cover the remaining areas of little conservation significance.

Plate 13. Diversity and age of a habitat are major criteria in the assessment of its wildlife value. This SSSI, on the eastern shore of Loch Lomond, comprises long-established mixed deciduous woodland supporting a wide variety of plant and animal species.

Inevitably, many planners will remain sceptical of the worth of a wildlife resource unless its value can be stated in more objective terms, and so some researchers have concentrated on the assessment of economic values. Broadly, there are three bases upon which the worth of wildlife to man can be appraised: as a reserve of native genetic material for breeding or for pollination or for scientific research and training; in terms of its opportunity cost where development potential is foregone in favour of retaining a wildlife community; and as a recreational and aesthetic resource for, for instance, natural history studies, wildfowling or for its general contribution to the character of a locality (16). The recreational component of wildlife value can be quantified on the basis of a demand curve derived for a site, which reflects the willingness of visitors to pay a cost (ie travel cost plus entrance charges) for the facilities. Everett (17) has combined such a model with an interview survey to elicit the relative importance which visitors ascribed to the presence of wildlife, and was thus able to obtain an estimate of the amenity value of wildlife in monetary terms.

It is now generally accepted that the importance of a wildlife habitat, from the educational and ecological point of view, attaches largely to the diversity of species which it supports and to the scarcity of those species. Its value is relative: for instance, arctic-alpine flora would be considered extremely valuable in Britain whereas in continental Europe, where distribution is less restricted, their conservation priority might be placed rather lower. To determine the relative value of a species in terms of rarity, it is first necessary to assume that the whole population will possess some collective worth (18). If total numbers are then assumed to be reduced the overall value also decreases, but not in direct proportion. For example, if a sizeable population of waterfowl were to decline by half, the birds would still be almost as noticeable and if conditions were favourable they would recover in numbers over a short period of years. It has been assumed that a reduction in numbers of 85 per cent would be required to halve the total value, with the relative value of each remaining individual rising steeply when more than 90 per cent of the stock had been lost (Figure 5.1).

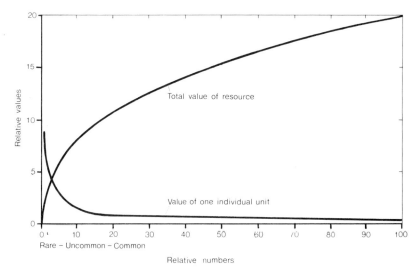

Figure 5.1   Relative values of individual members of an animal population in relation to total population size. (Source: Helliwell (16).)

Diversity of species also adds value to a site. Larger sites are, in general, more likely to support a more varied biota, albeit subject to a law of diminishing returns. Moreover, larger sites, because of their greater feeding areas, are more likely to sustain viable populations of rare species at the top of the ecological pyramid. Where site purchase is contemplated, it should prove possible to discern an optimum size in terms of relative value added per additional unit area.

Landscape

The objective description of landscape requires that an attribute which is appreciated essentially for its sense of wholeness should be recorded as a set of discrete elements, and that a quantitative score should be ascribed to something which is prized for its transcendence of measurable values. In addition, different professional groups such as planners, landscape architects, ecologists and artists will all possess varying concepts of landscape, and these will in turn differ from the predilections of the general public. The complex and variable combination of topographical, recreational, cultural, biological and scenic components must be reflected in evaluative techniques. A great deal of research has concentrated to this end and reference to a fuller discussion is recommended (19).

Assessments of individual sites likely to be affected by visually intrusive development normally involve the analysis of three main features: the extent of the area from which the site is visible, the quality of views into and out of the site, and the character of the area within which it is located. The first of these necessitates the determination of the zone of visual influence, which can usually be interpreted from a contoured map, taking into account installation size and site topography. Once this zone has been identified, a surveyor makes a subjective estimate of the magnitude of impact likely to be made upon it, by appraising the quality of views into the site from a number of apparently critical viewpoints. The effect of the development may then be determined in terms of the extent to which it substantially alters the character of the area, or detracts from its beauty, or is incompatible with the scale of relief (20).

The survey and evaluation of landscape on a regional basis is a rather more complex matter, and the search for a standardised method of appraisal has foundered upon a fundamental dichotomy. Hence, it so far remains unresolved whether landscape should be assessed in terms of the objective observation of specified physical components, or in terms of a subjective response on the part of the viewer to the 'overall' landscape. The former may be applied with greater consistency but the latter gives priority to the visual appreciation of fine scenery, which is the essential purpose of landscape survey.

Accordingly, there have evolved two categories of evaluative methodology, the first reflecting the degree to which an observer derives aesthetic satisfaction from a total scene, and the second, more analytical, in which individual geographical components are recorded and scored according to a pre-determined scale. The first, whilst theoretically preferable, requires considerable interpretive skill and entails total coverage in the field, whilst the second not only simplifies the level of understanding involved but also enables greater use to be made of maps and aerial photography. It follows that the choice of technique will normally rest upon the ability and experience of the surveyor and the time available to conduct the survey.

In one example of a simple, subjective technique, the sole criterion was the surveyors' personal consideration of what they deemed to be 'good' and 'bad' landscape (21). Each survey tract was thereby assigned to one of four categories, ranging from 'extremely attractive' landscape, worthy of preservation at all costs, through 'attractive' and 'average' views, to 'poor' landscape of no special value. Whilst such a classification might seem to be too informal to be of any value, it is, in essence, comparable to the manner in which most Areas of Outstanding Natural Beauty and of Great Landscape Value have been selected; moreover, it was found to be particularly straightforward to use and it produced a relatively high consistency of results between surveyors.

One of the best known objective methods is that devised by Linton, in which each grid square within the survey area is placed into two categories, one representing Land Use and the other Land Form (Table 5.4) (22). On the basis of these, composite scores reflecting overall landscape quality may be obtained. However, certain features of the technique may give rise to substantial variation in the results of individual surveyors, notably in the distinction between certain Land

LAND USE		LAND FORM	
Category	Score	Category	Score
Urban and industrialised	-5	Lowland	0
Blanket forest	-2	Low uplands	+2
Treeless farmland	+1	Plateau uplands	+3
Moorland	+3	Hill country	+5
Varied forest and moorland	+4	Bold hills	+6
Richly varied farmland	+5		
'Wild' landscapes	+6		

Table 5.4   Land use and land form scores in Linton's method of landscape assessment. (Source: Linton (22).)

Use classes and between different categories of Land Form, especially where the perceived boundary transects a grid square (23). Since the combined score relies heavily on Land Form this reduces the usefulness of the method for the purpose of management.

Recently, a major study of landscape evaluation has been carried out by Manchester University in an attempt to produce methods as objective and universal as possible in their application (24). Two techniques emerged as being appropriate to the needs of strategic planning, one requiring total field coverage by a team of surveyors and the other involving some use of laboratory analysis. In the first, two units are chosen and are then visited by all the surveyors, each of whom assigns to them a subjective value. Using these two values as a yardstick, the surveyors then proceed to assess the relative quality within each of the remaining units; afterwards, each observer gives a score, on the same scale, for the visual quality of the best and worst landscapes he believes to exist in the British Isles. These ranges are then normalised on a 1-100 scale and the field scores converted to this scale (for each surveyor); the mean score and standard deviation are then calculated for each survey unit (on the basis of all the surveyors' results). In the second method, estimates of weights are derived for specified landscape components on the basis of both a partial field survey and expert opinions: the actual weights to be used are then obtained by a regression analysis of these two sets of data. The assessments of quality derived by field survey are then related to identifiable landscape features by principal components analysis, and a combination of the weights and principal components may then be used to predict values for those units not surveyed in the field.

Despite the increasing rigour with which techniques of landscape appraisal are being developed, little consensus has so far been achieved about any one approach; indeed, no matter how objective and sophisticated evaluative procedures might become, it is probable that there will always be a place for the simple, subjective appraisal. The assessment of landscape will depend to such an extent on the preference of the user, the purposes for which it is being carried out, and the characteristic scenery of a particular region, that it is improbable that a single standard survey technique will ever find universal acceptance.

Water and minerals

Mineral reserves which are accessible by deep mining, whilst they may present problems of subsidence and visually intrusive pithead machinery, remain relatively unaffected by surface development. Deposits which are to be quarried or recovered by opencast methods — such as sandstone and limestone (including sand and gravel), roadstone, peat, and certain measures of coal and metal ores — must, however, be protected from urban encroachment.

The value of a mineral deposit is related to its quality or grade, which may vary greatly over short distances, its thickness, and the depth under which it lies. In addition, economic factors, such as the current and forecast market values of the mineral in question, its local availability, and the costs of importing it from neighbouring regions will affect its conservation priority.

The Geological Survey of Great Britain has concentrated its mapping of solid and drift geology upon the most densely populated areas of the country and areas of known mineral wealth; information on economic reserves of minerals should therefore be available for most planning purposes (25). On the basis of the cartographic and bore hole data which this Survey furnishes, interpretive techniques may be applied to obtain estimates of the volume of deposits. When such estimates are allied to an economic appraisal of a reserve, a potential market value for it may be obtained, which may serve as an indicator of its relative conservation priority. The likelihood of a reserve being exploited in the future should also take into account amenity considerations, giving regard to the scale of intrusion which workings would entail in the light of the surrounding topography and the likely effectiveness of artificial landscaping or screening.

Reserves of potable water supply may be similarly affected by urban developments and other pressures which modify the land surface. A high degree of restriction on access in the gathering grounds of reservoirs has traditionally been enforced by water supply authorities, but there is now a greater realisation that urbanisation may markedly affect the manner in which aquifers and watercourses are recharged. Since these sources may themselves be used for abstraction, or may interconnect with reservoirs and thus be partially responsible for their replenishment, consideration must be given to any development which could affect the volume of purity of water entering them.

The problems of river pollution, both in amenity and economic terms, have long been subject to assessment. Watercourses in Britain are graded into four classes of relative quality, namely those which are unpolluted or have recovered from pollution (Class 1), those of doubtful quality and needing improvement (Class 2), those of poor quality requiring improvement as a matter of some urgency (Class 3), and those which are grossly polluted (Class 4). About 10·5 per cent of river miles in England and Wales fall into classes 3 and 4. Other factors in addition to water purity may also influence the value of a river to nature conservation. These include the degree of human disturbance, the variety and number of micro-habitats (niches) along the river bank, and the nature of the river bed; however, the relative importance of these various features is as yet poorly understood.

Countryside recreation

An analogy has been drawn between agricultural capability assessments, which establish relationships between land properties and crop requirements in order to indicate optimum farming practices, and recreational capability evaluations in which a relationship is sought between land properties and user requirements, with the object of maximising user satisfaction. A simple classification of optimum criteria for the most popular land- and water-based pursuits has been proposed as suitable for use at the sub-regional level (Table 5.5) (26).

Camping: caravanning: picnicking:	All countryside within 400 m of a metalled road
Pony-trekking:	All upland areas above 300 m in altitude and with rights of way or established footpaths and bridleways
Walking and hiking:	All upland areas above 450 m, with rights of way or established footpaths and bridleways
Game shooting:	All areas assessed as shootings on valuation rolls
Rock climbing:	All cliff faces over 30 m in height
Skiing:	Available relief of over 280 m, with an average snow-holding period of more than three months

Table 5.5 Factors in recreation classification. (Adapted from Coppock (26).)

Certain problems are evident in this approach, however. In practice, once again we find that many user requirements are unquantifiable or intangible and are subject to the vagaries of human taste, whilst tests designated to measure user satisfaction tend to be circular in their reasoning. An attempt to achieve a more reliable procedure has been made by Johnstone and Tivy (27), who considered in detail the recreation potential of lochside site-types. Transects from the road to the lochshore were recorded to provide objective descriptions of the site in terms of shore features, accessibility, slope and vegetation type. Once this basic inventory had been compiled, the shore and backshore could be evaluated in terms of their suitability for selected recreational pursuits, according to the degree to which the site approximated to a set of pre-determined characteristics. For instance, optimum conditions for mobile homes, trailers and caravans etc, on the backshore included: a south-facing aspect, negligible flood hazard, gentle gradient, rapid drainage and wide parking areas. Each of those attributes could then be scored, and weighted to reflect the ease with which low level management could alleviate the particular site limitation. The capability class to which a site was allocated would then be determined by summing the weighted scores. The technique is, however, laborious, and it is unclear how suitable it might prove for sub-regional surveys of general recreational potential.

Plate 14. Rural planning must seek to ensure the most rational single – or multiple – use of the countryside. The Goyt Valley, in the Peak District National Park, exemplifies a sensitively managed combination of farming, forestry, water catchment and recreation.

Optimisation Techniques

Strategic appraisal

Rural areas of high quality are usually treated not as rich resources but as constraints to development in most statutory plans. Consequently, there is a predominantly negative and restrictive approach to country-side planning. Furthermore, such specific resources as agriculture and wildlife tend to be treated in a topical and isolated manner; little attempt is made to integrate them into a general sub-regional strategy. However, some examples of how matters could be improved are emerging.

One technique which has been employed to ensure a more even treatment of rural and urban resources is potential surface analysis (PSA). This entails the division of a region into zones, on the basis of a regular grid, and the assignation to these of scores which represent the relative potential of each of a range of selected features. Thus, if the suitability of areas for industrial development is the attribute to be determined, a high score on the index representing accessibility would enhance the zone's potential, whilst a high score for agricultural capability would depress its attractiveness. The scores for each land feature are usually weighted to reflect its relative importance; these are then combined for each zone, in order to derive a composite score which represents the overall potential for development or con-servation (28).

A modification of PSA, termed 'land capability analysis', has been used as a basis for the assessment of rural development in the North Yorkshire Moors (29). Evaluations of land capability were obtained for agriculture, forestry, wildlife and recreation, the relative quality of each being expressed on a five point scale which was assumed to be directly comparable for each use under consideration. Transparent maps were then compiled showing the scores for each use, which when overlaid, threw into relief the one or more optimal uses for each unit of land. The next highest score use was then obtained for each unit, giving a secondary activity and thereby introducing an element of multiple use into the final land pattern. The weighting of different activities in order to reflect a range of policy options led to the production of a set of alternative scenarios.

Amongst the advantages which have been claimed for PSA are: its use of indices which provide a continuous evaluative scale rather than simply an arbitrary cut-off into good and bad areas; the systematic and explicit weighting of factors which it permits, the supposed compa-rability of estimates of quality which it applies to each resource; and its effectiveness in identifying areas of conflict which can then be studied

	Present land use pattern	Optimum patterns: maximising timber		Optimum patterns: maximising meat		Optimum patterns: maximising food energy	
	Totals	Totals	% of present totals	Totals	% of present totals	Totals	% of present totals
Timber (tonnes)	331 000	450 510	36	331 000	0	331 320	1
Meat (tonnes)	81 800	81 800	0	83 904	3	81 800	0
Food energy (terajoules)	3 470	3 470	0	3 470	0	4 143	19
Milk (tonnes)	843 000	918 471	9	910 390	8	1 144 910	36
Wool (tonnes)	2 700	2 700	0	2 700	0	2 700	0
Recreation	332 000	332 063	0	332 130	0	332 048	0
Ecological value	257 000	287 050	0	287 096	0	287 081	0
Labour input (millions SMD)	3.60	3.61	0	3.60	0	3.86	7
Energy input (terajoules)	9 600	9 951	3	10 006	4	10 379	8

Table 5.6   Outputs and inputs from Cumbria under present and proposed land use configurations. (Source: Cumbria County Council/Lake District Special Planning Board (30).)

in greater depth. However, the technique also possesses certain weaknesses including the omission of economic or social factors (although those may be implicit in the weightings), the risk of double counting (overweighting certain attributes by the inclusion of overlapping indices), and its greater aptitude for identifying development potential rather than conservation need.

A possible alternative, which perhaps incorporates a greater element of dynamism, is the use of budgeting procedures for human ecosystems (see also Ch. 2). These enable land use patterns to be derived for these which encourage maximum efficiency of energy conversion in key production areas and enable conservation resources to be protected elsewhere. A pioneering approach in this field has been taken by Cumbria County Council in conjunction with the Institute of Terrestial Ecology (30). Using a zonal classification of the county, designed to reflect the intrinsic productivity of each parcel of land, possible departures from the current land use pattern were examined in the light of their changed landscape impacts, labour requirements and energy subsidies. In particular, the consequences of pursuing three alternative policy options — which sought to maximise timber production, meat production and food energy respectively — were explored. A mathematical optimising technique, 'linear programming', was employed to determine the best levels of production of each commodity in relation to each policy objective. Linear programming also possesses the important property of enabling optimisation to be made subject to constraints; it was therefore possible to ensure that no reduction in recreational and ecological value occurred, and there was especial regard to those zones which formed part of a National Park, an Area of Outstanding Natural Beauty or common land. Perhaps somewhat surprisingly, substantial increases in commercial production proved possible with no apparent reduction in ecological or recreational potential (Table 5.6).

These conclusions would seem to suggest that there is a great deal of unrealised potential and unnecessary conflict which, by ingenuity and perseverence in the use of more complete assessments, could be realised and resolved.

Site appraisal

Until relatively recently the scale of industrial activity and infrastructure, especially in rural areas, has been rather modest and ordinary development control procedures have proved adequate. However, the increasing tendency for such structures to make unprecedented demands upon an area, either by virute of their size or the nature of their processes, has made necessary the introduction of more sophisticated methods of assessment. Accordingly, techniques of environmental impact analysis (EIA) were pioneered in the USA to enable more systematic investigations of major proposals in terms of their social, economic and ecologic consequences (31). *The National Environmental Policy Act, 1969*, then made compulsory the preparation of an Environmental Impact Statement (EIS) for any project undertaken or funded by federal agencies; a Council on Environmental Quality was established to advise on the format and content of statements, as well as to scrutinise those submitted.

Several EEC countries subsequently adopted some form of EIA procedure, although in all cases this has been grafted onto the existing planning machinery, whereas in America a completely new appraisal system had to be introduced.

The Federal Government of West Germany, in its Environmental Programme of 1971, determined that an 'examination of environmental compatibility', conducted by the ministry concerned, would be an essential factor in the preparation of all measures by federal authorities. The French *Protection of Nature Act 1976* provided that prior to the commencement of significant public works, or private projects requiring public authorisation, an impact assessment must be carried out in accordance with specific requirements of coverage, content and public participation. In Eire, the *Local Government (Planning and Development) Act 1976* empowers (but does not oblige), the relevant minister to require the preparation of an EIS for any project having a value in excess of £5m. Similar proposals are under consideration in the Netherlands (32). In the UK a number of *ad hoc* EIAs have been carried out and, in August 1979, the government stipulated, for the first time, a form of mandatory EIA as a prerequisite when considering any future developments associated with petrochemicals at Moss Morran, Fife.

It is current opinion within the EEC that EIAs are most suited to: industrial investments such as metallurgical projects, metal working, chemical, textile and leather industries; major transport installations, notably motorways, airports and harbours; and electricity power lines. However, the procedures would also be appropriate to other installations associated with hazardous or noxious wastes, to major urban developments and even to certain types of agricultural improvements (33).

In the UK, the precise form and content of an EIA and the choice of body responsible for its preparation (district or county/regional planning department, developer, independent consultants, or Planning Inquiry Commission, for instance) are issues which have still not been finally settled. However, both the Department of the Environment and the Scottish Development Department have collaborated in the preparation of a standardised version which has the advantage of being compatible with current development control procedures (34).

The approach adopted incorporates many features common to the variety of analytical formats so far suggested for EIA. It pays regard to a variety of development effects, most notably the visual impacts of scale, the implications for plant and animal life and the consequences of development on the livelihoods of people. It commences with a 'baseline study' which furnishes a detailed description of the topographic, ecological and socio-economic composition of the proposed site. It goes on to describe the proposals in terms of their construction time, potential hazards, consumption of resources, end products and waste materials. Finally, it includes an analysis of the probable primary and secondary impacts of development on the environment during both the construction and operational phases, these being summarised in a matrix of interactions (Figure 5.2). In order to elicit answers, a checklist of questions must be completed by the developer ('project specification report') and this provides the factual basis for the identification and forecasting of impacts.

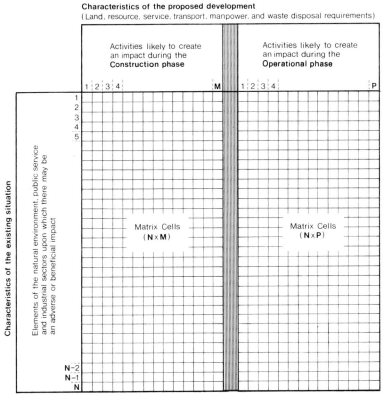

Figure 5.2   Generalised format of the interactions matrix, at the heart of the method of impact analysis devised for the Department of Environment and the Scottish Development Department by Aberdeen University. (Adapted from Clark et al (34).)

Clearly, EIA could ensure a more thorough and effective appraisal of proposals than would otherwise tend to occur and would enable the general public to contribute, in a more informed and effective manner, to the determination of decisions which significantly affect their own environment. Moreover, EIA could foster a broader appreciation of the irretrievable and irreversible commitments of resources which tend to be involved in securing short term benefits. Perhaps most importantly, this technique could provide a greater assurance that environmental matters were taken into account than if each developer considered them on a purely internal basis — an observation especially true of statutory undertakers, who often escape the scrutiny of 'normal' planning controls (35).

Despite these advantages, it is only fair to point out that critics of the EIA system consider it little more than a device for objectors to delay the planning process and to cause many socially desirable projects to be cancelled or postponed indefinitely.

Economic appraisal

An assessment of the intrinsic productive value of a unit of land can only provide a part of the data on which planning decisions must be based. As we noted earlier, a resource is essentially a cultural concept having a relevance only within the context of the human economy: an underlying assumption must therefore be that, whilst ecological considerations impose the ultimate limits on land use and are therefore paramount, the actual nature of resource exploitation must satisfy economic criteria also.

The environment presents us with a range of alternatives, between short-term gain and long-term stability, and it is often to the planner that the task of evaluating each course of action falls. Decisions in the private sector usually rely on a comparison of the quantifiable economic costs and benefits associated with investment proposals, but the techniques used for this purpose often transpose rather imperfectly for use in the public sector. The distributional outcome of a proposal through time (among present or future generations) or in space (over people and places), and 'intangible' costs and benefits, not subject to the mechanism of the market, and to which no monetary value can be ascribed, are not accounted for in such methods.

The use of *cost-benefit analysis* in environmental planning must seek to determine that combination of development proposals which produces the greatest net benefit to the community (36). In so doing, it presupposes that we are able to evaluate various different kinds of social cost, including: those directly measurable and associated with the control, avoidance or consequences of environmental damage; the indirectly measurable costs, for which market values can be inferred from suitable proxies; and those unquantifiable costs where no values are readily attributable, such as the costs of illness or premature death. A number of surrogate variables have been advanced for the estimation of some of these: simple measures, such as the change in property values following the development of adjacent land, or composite 'annoyance indices' based on such factors as traffic noise or congestion are common (37). On occasion, however, it may prove desirable to employ a more sophisticated technique to elicit environmental expectations; one method, for instance, uses subject responses to photographs depicting varying levels of amenity provision as a basis for determining the kind of trade-offs between environmental cost and benefits which will prove socially acceptable (38).

A major difficulty in determining the ratio of costs to benefits associated with the exploitation of natural resources lies in the economic appraisal of their worth in relation to alternative land uses. Thus, the true value of a natural resource lies in the fact that, if properly managed, it will continue to furnish the needs of future generations. If it is proposed to build an office block on prime farmland, the revenue from the commercial use may be very high but relatively short-lived, whereas the return from agriculture may be comparatively low, but will accrue in perpetuity. The planner's dilemma is thus how to draw a fair comparison between long- and short-term benefits.

The standard approach to this problem is to discount the future returns from an activity to the present, so that a *net present value* (NPV) may be obtained. This assumes that future benefits may be aggregated, but that the returns from future years must be reduced — *ie* discounted at a selected percentage rate — to allow for the effects of depreciation and inflation. Clearly, if a high discount rate is selected, the benefits from future years will quickly reduce to a negligible amount. If a ten per cent rate is chosen, and this is a common Treasury preference, returns anticipated beyond the twentieth year will be so small as to make little difference to NPV. Consequently, if a resource has an active lifetime beyond twenty years, it may be that its true value will be seriously underestimated, especially — and this will almost certainly be true of energy, food and minerals — if its value is likely to increase substantially in real terms in the future. The selection of a high discount rate may, for instance, encourage the excessive uptake of farmland for urban development or very rapid rates of depletion of oil reserves.

A similar investment problem arises in the uplands, where land managers must often decide between hill farming (which yields an annual return) and forestry (in which capital will be locked up for about fifty years). Again, the economic comparison must be made by discounting the eventual revenue from timber to obtain a NPV, but the high rates of discount preferred by the Treasury make these revenues appear poor in relation to hill farming, and this accounts for the rather conservative planting targets of the Forestry Commission (Figure 5.3). The Centre for Agricultural Strategy has recently argued for a greatly accelerated programme of afforestation, on the grounds that the future economic benefits of a home timber industry have been seriously underestimated (39).

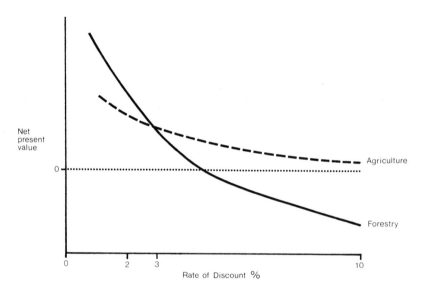

Figure 5.3   Comparative economic returns from agriculture and forestry on moderately good hill land. (Adapted from Johnson et al).)

One of the major problems in the use of cost-benefit analysis for a particular project is that it was designed to concentrate somewhat exclusively on the criterion of cost efficiency, and as a consequence there is no real procedure for accounting for the multiple objectives which public sector programmes are normally intended to achieve.

An alternative approach is afforded by the *goals achievement matrix*, in which proposals are appraised by reference to their relative success in satisfying a spectrum of selected criteria. Hill and Alterman (40) have described a simplified version of this technique, termed the *criteria evaluation matrix*, by which an attempt was made to determine the comparative merits of two power stations sites in Israel. The environmental, social and economic effects likely to be caused by each alternative were studied by specialist teams who were able to derive a relative ranking of the two sites on the basis of pre-determined criteria (Table 5.7).

Criteria	Zarka Site	Hadera Site
Contribution to sub-regional development objectives	--	++
Accordance with land use zoning principles	+	-
Ecological impact	--	++
Pollution levels	-	--
Economic appraisal (cost-benefit ratios)	ratios similar	
Legal-administrative simplicity and desirability	-/=	+/=
Public perception	-	+

Table 5.7   The use of a criteria evaluation matrix in the determination of the environmental desirability of alternative schemes. (Adapted from Hill & Alterman (40).)

'Input-output analysis' has been widely advocated as a tool for the evaluation of regional or industrial economic performance, for it has the advantage of operating upon a relatively closed system and thus ensures that all inter-regional or inter-industry flows of raw materials and finished products are accounted for (41). The framework ensures that a trading balance must be struck between two sectors and that the ultimate requirements of consumers must be equated with the pattern and level of industrial activity necessary to satisfy this final demand. However, it could equally well be contended that the performance of industry has been considered in isolation from its effects upon the ultimate 'source' and 'sink' of all comodities, the environment. In terms of the economy of nature, therefore, the accounting framework cannot be considered truly 'closed', and in order to rectify this deficiency, techniques of *ecologic-economic input-output analysis* have been developed (*cf* 'space-ship economy', Chapter 6).

Just as the basic model attempts to identify monetary flows and balance of payments, so the ecological model seeks to establish user-environment flows of raw materials and waste products and to achieve a 'materials balance' (42). This is effected by regarding the environment as an additional industry in the input-output framework and it thus becomes possible to estimate the ecological impact of the level of industrial activity required to sustain a specified pattern of consumer demand.

One application of this method was used to determine the respective merits of two alternative sites for a marina development at Plymouth Bay, Massachusetts (43). Each of the main operations, such as

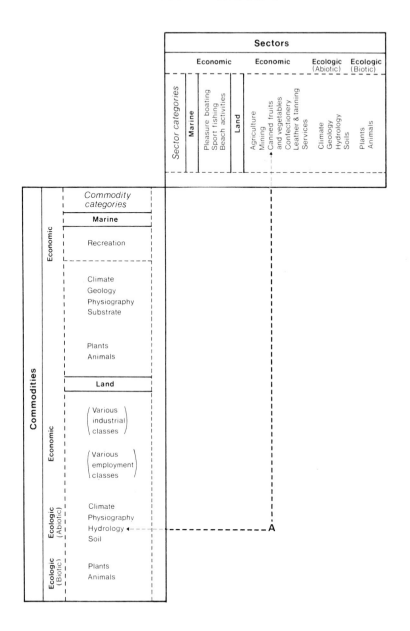

Figure 5.4   Simplified version of ecologic-economic input-output framework for the analysis of marine development proposals. Where appropriate, coefficients based on forecast levels of interaction between sectors and commodities are entered into matrix cells, eg, the value in cell 'A' reflects the water intake requirements of the existing canning sector. (Adapted from Isard (43).)

dredging, shore protection, provision of sewage disposal facilities and construction of breakwaters, was assumed to affect a certain natural resource and to produce a desired recreational commodity (Figure 5.4). To this end, in addition to measuring the comparative economic costs and recreational and employment benefits, a simultaneous assessment was also made of the ecological impacts of each alternative. Although it could be argued that the key problem of ascribing values to ecological commodities was not really overcome, the exercise nevertheless did have unquestionable merit from the point of view of theoretical ecology: it recognised the dynamic nature of ecosystem flows, and in this respect it represented a major advance on conventional frameworks which tend to treat the environment merely as a static backcloth to man's activities.

NOTES

(1) M. Gane (1978) 'The balance of rural land use' in J. G. Hawkes (Ed.) *Conservation and Agriculture*, Duckworth.
(2) One application of land assessment for planning purposes is described in: Joy Tivy (1980) 'Information for regional land use options', *Town Planning Review*, 51.
     A useful general review of regional appraisal techniques is contained in M. F. Thomas and J. T. Coppock (Eds.) (1980) *Land Assessment in Scotland*, Aberdeen University Press.
(3) J. S. Bibby and D. Mackney (1969) *Land Use Capability Classification*, Soil Survey Tech. Monogr. No. 1 Rothamstead Exptl. Stn.
(4) M. A. B. Boddington (1978) *The Classification of Agricultural Land in England and Wales: a critique*. Homer, Oxford, Rural Planning Services Ltd.
(5) Highlands and Islands Development Board (1973) *Island of Mull – survey and proposals for development*, HIDB Spec. Rep. No. 10, Inverness, HIDB.
(6) Anne Dennis (1976) 'Agricultural land classification in England and Wales', *Planner*, 62. But see also the critique of this in Boddington *op. cit.* (4).
(7) A. W. Gilg (1975) 'Development control and agricultural land quality', *Town and Country Planning*, 43.
(8) D. R. Johnston, A. J. Grayson and R. T. Bradley (1967) *Forest Planning*, London, Faber.
(9) Veronica Burbridge (1971) 'Methods of evaluating rural resources: the Canadian experience', *RTPI Journal*, 57.
(10) R. D. Toleman (1974) 'Land capability classification in the Forestry Commission', in MAFF, *Land Capability Classification*, Tech. Bull. No. 30, London, HMSO.
(11) D. G. Pyatt, D. Harrison and A. S. Ford (1969) *Guide to Site Types in Forests of North and Mid-Wales*, For. Comm. For. Rec. No. 69.
(12) D. Statham (1972) 'Natural resources in the uplands: capability analysis in the North York Moors', *RTPI Journal*, 58.
(13) See, for example, D. A. Ratcliffe (1971) 'Criteria for the selection of nature reserves', *Adv. Sci.*, 27.
(14) C. R. Tubbs and J. W. Blackwood (1971) 'Ecological evaluation for planning purposes', *Biol. Conserv.*, 3.
(15) D. A. Ratcliffe (Ed.) (1977) *A Nature Conservation Review*, 2 vols., Cambridge University Press, for the Nature Conservancy Council. A useful account of

this work is given in G. Smart (1978) 'Nature conservation and planning', *Town Planning Review*, 49.

(16)   D. Helliwell (1973) 'Priorities and values in nature conservation', *Journal of Environmental Management*, 1.

(17)   R. D. Everett (1978) 'The monetary value of the recreational benefits of wildlife', *Journal of Environmental Management*, 6.

(18)   D. Helliwell *op. cit.*

(19)   See, for instance: E. Penning-Rowsell (1973) *Alternative Approaches to Landscape Appraisal and Evaluation*, Enfield, Middlesex Poly. Plann. Res. Gp. Rep. No. 11; D. G. Robinson (1976) 'Rural Landscape' in G. E. Cherry (Ed.) *Rural Planning Problems*; Aylesbury, Leonard Hill; and I. C. Laurie, D. G. Robinson, A. L. Traill and J. F. Wager (1976) *Landscape Evaluation Research Project*, Report of the LERP to the Countryside Commission for England and Wales, Manchester University Press.

(20)   R. Hepplethwaite (1973) 'Landscape assessment and classification techniques' in D. Lovejoy (Ed.) *Land Use and Landscape Planning*, Aylesbury, Leonard Hill.

(21)   E. C. Penning-Rowsell and D. L. Hardy (1973) 'Landscape evaluation and planning policy: a comparative survey in the Wye Valley AONB', *Regional Studies, 7*.

(22)   D. Linton (1968) 'The assessment of scenery as a natural resource', *Scott. Geogr. Mag.*, 84.

(23)   A. W. Gilg (1976) 'Assessing scenery as a natural resource: causes of variation in Linton's method', *Scott. Geogr. Mag.*, 92.

(24)   I. C. Laurie, *et al* (1976) *op. cit.* (19).

(25)   A useful critique of the Geological Survey and other sources is provided by: D. K. Chester (1980) 'The evaluation of Scottish sand and gravel resources', *Scottish Geographical Magazine*, 96.

(26)   J. T. Coppock, B. S. Duffield and D. Sewell (1971) 'Classification and analysis of recreation resources' in P. Lavery (Ed.) *Recreational Geography*, David and Charles.

(27)   May Johnstone and Joy Tivy (1980) 'Assessment of the physical capability of land for rural recreation' in Royal Scottish Geographical Society, *Symposium on Land Assessment in Scotland*, Aberdeen University Press.

(28)   A general discussion of PSA is given by J. Zetter (1974) 'The application of potential surface analysis to rural planning', *The Planner*, 60.

(29)   D. Statham, *op. cit.* (12).

(30)   Cumbria County Council/Lake District Special Planning Board/Institute of Terrestrial Ecology (1978) *An Ecological Survey of Cumbria*, Working Paper 4.

(31)   J. Catlow and C. G. Thirlwell (1977) *Environmental Impact Analysis*, Department of the Environment Research Paper No 11, London, HMSO.

(32)   N. Lee and C. Wood (1978) 'Environmental Impact Assessment of projects in EEC countries', *Journal of Environmental Management*, 6.

(33)   T. O'Riordan and R. D. Hey (Eds.) (1976) *Environmental Impact Analysis,* London, Saxon House.

(34)   B. D. Clark, K. Chapman, R. Bisset and P. Walthern (1976) *Assessment of Major Industrial Applications*: a manual, Department of the Environment, Research Report No 13, London, HMSO.

(35)   Anne Beer (1977) 'Environmental Impact Analysis — a review article', *Town Planning Review*, 48.

(36)   For a general account of benefit-cost analysis and its role in planning, see, for instance, G. A. Chadwick (1971) *A Systems View of Planning*, Oxford, Pergamon, pp. 263-67.

(37) Examples of 'annoyance indices' and related measures are to be found in: Greater London Council (1974) *London's Environment* — 1st, Report of the Environmental and Pollution Control Group; County Hall, London and in A. A. Walters (1975) *Noise and Prices*, Oxford, Clarendon Press.

(38) G. Hoinville (1971) 'Evaluating community preferences', *Environment and Planning*, 3.

(39) Centre for Agricultural Strategy (1980) *Strategy for the UK Forest Industry*, University of Reading.

(40) M. Hill and R. Alterman (1974) 'Power plant site evaluation: the case of the Sharon plant in Israel', *Jour. Eviron. Mgnt.*, 2. Hill had previously developed the idea of 'goals achievement' in: M. Hill (1968) 'A goals achievement matrix for evaluating alternative plans', *Journal of the American Institute of Planning*, 34.

(41) Useful reviews of input-output analysis may be found in: J. R. Meyer (1968) 'Regional economics — a survey' in L. Needleman (Ed.) *Regional Analysis*, Harmondsworth, Penguin; and in G. Chadwick *op. cit.* (19), pp. 226-29.

(42) For a development of materials balance equations see R. H. Pantell (1976) *Techniques of Environmental Systems Analysis*, NY/London, Wiley.

(43) W. Isard (1972) *Ecologic-Economic Analysis for Regional Development: some initial explorations with particular reference to recreational resource use and environmental planning*, New York, Free Press. For an extension of this approach to the British economy, see P. A. Victor (1972) *Pollution: economics and environment*, London, Allen and Unwin.

6 Governance of the Environment

According to one American economist, governments have tended to view the national economy as a production process in which inputs mysteriously appear at one end and outputs disappear at the other. The consequence of this has been a 'cowboy economy', characterised by flamboyance and wastage, in which no account is taken of ecological constraints. It would be more responsible, he suggests, to treat the environment as a closed system in which there is parity between its role as supplier of finite goods and resources and its capacity to assimilate both desired and waste products. This would effectively be a 'space-ship economy', in which, like astronauts, we were obliged to protect the quality of our living space and to conserve our supplies (1).

In trying to convince governments to adopt a more comprehensive approach to national wellbeing a major problem is the way national performance, in regard to economic growth, is measured. Traditionally this is reflected in the 'Gross National Product' which represents national output of goods and services at market prices — a poor indicator (to environmentalists) of true prosperity. Not only does the market fail to place adequate values on collective or intangible goods, such as visual beauty or air and water quality, but it takes into account only those items and services which can be sold at a profit — 'goods' — and ignores undesirable side effects — 'bads' — such as pollution, for which no exchange value exists. More fundamentally, many critics have questioned the pursuit of permanent growth itself, although the concept of 'growth' in human societies is admittedly complex and the adoption of no-growth goals begs many questions (2). A broader concept, a 'Measure of Economic Welfare', has been put forward by some commentators as an alternative to replace the GNP (3).

An environmentally conscious government would respond to a variety of welfare indicators and act according to the precepts of the spaceship economy. In particular, it would find ways to manage both ends of the production function: it would ensure a continued supply of raw materials by monitoring responsible consumption of natural

125

resources and conserve environmental quality by restricting pollution levels. Since private operators consider only short term profitability it falls to a responsible public sector to satisfy the broader social good. Two examples serve to illustrate the point.

Where ownership of/control over a resource is fragmented or ill-defined, market exchanges become inefficient or impossible. When all who have a claim to them scramble for the usage of such collective goods as petroleum pools, ocean fisheries or recreation facilities the result is likely to be excessive rates of depletion; this form of depredation is referred to as the 'tragedy of the commons' after the fate of common grazing land in earlier times (4). In the second example, a manufacturer whose plant is a cause of pollution will be unconcerned about the nuisance to people living in the vicinity since that would be external to the aims of his enterprise. Such undesirable neighbourhood effects are termed 'negative externalities' by economists (planners usually refer to 'disamenity' or 'nuisance'). Such problems occur because the natural environment is seen as being external to commercial operations; within a spaceship economy, conversely, no effect could be truly external since all activities are encapsulated within overall ecological limits.

The two major responsibilities of the public sector are, therefore, to ensure the wise use of resources by exercising an influence over land use distribution and to 'internalise' externalities (*ie* to maximise amenity or minimise disamenity) by ensuring that the general public interest is taken into account when potentially controversial developments are proposed. The British town planning system is a good example of this kind of regulatory activity although it is only one of the many agencies through which government may influence the management of the environment.

Economic Determinants of Administrative Action

Pollution control

The control of pollution in Britain has traditionally been resolved in relation to local conditions, especially those of residential amenity and the capacity of individual receptor environments to accept pollutants. Increasingly, however, overseas governments have begun to favour uniform national standards and this certainly appears to be the approach being promulgated by the EEC. With the advent of a supra-national dimension to environmental quality it would seem appropriate now to consider the general principles which govern the optimum levels of pollution control.

In an ideal world we would seek to eliminate pollution totally but, in

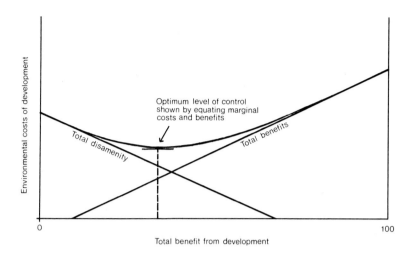

Figure 6.1 Determination of the optimum level of pollution control. (Source: Wood (19).)

practice, this would prove impracticably costly. The economist's solution is to locate the optimum trade-off between the cost of abating pollution and the benefits of environmental quality accruing to the public. In Figure 6.1 the 'best' level of control occurs at the point where the combined damage and control costs are at a minimum. The problem then arises as to who should cover the bill for the measures necessary to achieve this solution — the producer who creates the problem or the public who will enjoy improved standards? In Britain, the approach has been to oblige the former to absorb the costs — the 'polluter pays' principle. This concept is illustrated in Figure 6.2 in which DD' is the market demand curve and SS' the market supply curve for a given commodity, K. The production of the commodity is assumed to cause a constant amount of pollution per unit output, the value of the latter to the producer being equivalent to £ab per unit output of k. Conventionally, it would be argued that market equilibrium had been reached where an amount ON of k was exchanged at price OP. However, if the producer were forced to pay compensation to those adversely affected, or to install plant to eliminate the pollution, an alternative optimum would arise at ON_1 (5).

It has been suggested that taxes might be levied in order to oblige polluters to account for the external diseconomies they cause: in particular, a 'residuals tax' would provide an incentive to incorporate abatement measures up to the point at which their cost exceeded the social damage incurred. Thus, a tax of £ab on the sale of k could be imposed, thereby raising the supply curve of the industry to S"S''' and

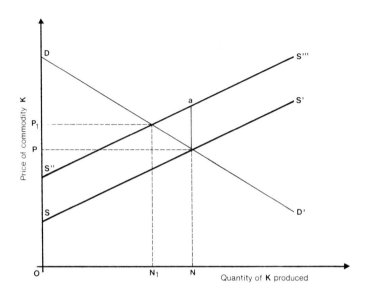

Figure 6.2   Supply-demand curve for pollution control. (Source: Victor (5).)

automatically reducing the quantity of k sold, to ON_1, at a market price inclusive of tax OP_1.

Alternatively, the creation of unnecessary waste might be curbed by reducing the rate of premature obsolescence via a 'throughput tax', in which a levy would be charged on the original producer of a commodity, the level of which was adjusted to reflect the social costs of the most environmentally harmful manner of disposal. A part of the tax could then be refunded in proportion to the extent to which the final disposal of the product was delayed.

Resource management

We have noted that resources may be broadly categorised according to whether they are renewable or non-renewable. An example of a renewable resource is provided by food production, in which we may assume exploitation (*ie* the amount of land brought into cultivation) to increase exponentially over time until the physical limit (C) is reached (Figure 6.3) (6). In theory, growth should be halted at this point, and the maximum yield consistent with the maintenance of future productivity, or maximum sustainable yield (MSY), adhered to thereafter. However, if consumers have become accustomed to a rising standard of living, there will be a tendency to exploit beyond the MSY, even though this may result in the permanent impairment of productivity.

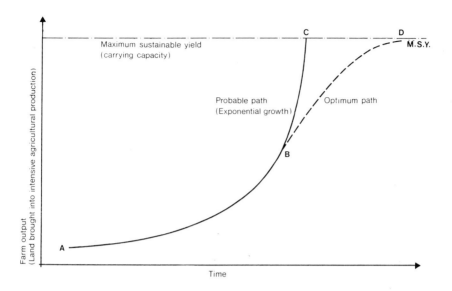

Figure 6.3 The limits to exploitation of a renewable resource: food production. (Adapted from Lecomber (6).)

Further, due to the exponential nature of growth of consumption, this limit is approached with rapidity in the later stages, providing policy makers little time to adjust, and it would naturally be preferable to anticipate this limitation, attempting to follow the path ABD. In practice, technological innovation and changes in dietary habits will also tend to raise the MSY but the evidence suggests that this increase could not be sustained indefinitely.

The response of governments to agriculture has thus been to encourage increases in productivity by providing advice and incentives to farmers. In Britain, where an efficient agricultural sector was long viewed as an optional adjunct to our manufacturing and mercantile activities, little or no action was taken by the Government until after World War II, and the home farming industry had foundered in a depressed state for many decades. The strategic vulnerability which this posed for us in times of siege, however, led to legislative action in the form of the *Agriculture Act 1947*, which provided for a comprehensive package of grants and subsidies.

Immediately after World War II the emphasis was on increased production, and the lynch-pin of this policy was a system of deficiency payments made to farmers to ensure that the market prices which they obtained for their products were raised to guaranteed levels. By the mid-1950's, however, Britain's strategic position had markedly improved, and, with cheap foodstuffs flooding in from abroad, greater attention could be paid to improving the efficiency of farms. Consequently, during the next decade or so, grants were introduced to encourage the investment of capital and rationalisation of farm structure; there was an increase in the proportion of large, highly mechanised units and a decrease in agricultural labourers — resulting in their exodus from the countryside. Since joining the EEC, adherence to the Common Agricultural Policy has entailed phasing out deficiency payments and replacing them with interventionist policies designed to maintain adequate farmgate prices, although most of the improvement grants previously available to British farmers have been replaced by approximate European equivalents. In addition, the disappearance of traditional overseas supplies of cheap food has led the British Government to exhort farmers to increase the level of net self-sufficiency in agricultural produce from 50 to 60 per cent in the medium term, and to 75 per cent by the turn of the century (7).

In making improvements, the farmer is faced with the prospect of substituting between his inputs of land, capital and manpower, leading to capital-intensification; and also of varying his product combination, resulting in a tendency towards monoculture. Where, then, the nature of an enterprise renders it susceptible to economies of scale, for instance in arable farming, the high cost of the indivisible inputs (especially machinery) tends to lead to increasing farm size and to concentration on a single combination of crops (8).

On occasion, farm layout may detract from the advantages of size, and, in particular, fragmentation may present a considerable obstacle to efficient operation. A number of government initiatives have been rather unsuccessful in achieving greater consolidation; in the uplands, where this problem is most acute, Rural Development Boards were established under the Agriculture Act 1967, but these were abruptly disbanded following a change of Government in 1970.

Land values and rents should, theoretically, reflect both the productivity of the land and the opportunity costs of using it for agriculture, rather than for industrial, residential or commercial development. But, in practice, a variety of institutional factors may grossly inflate its true value and make it attractive for investment purposes: recent research suggests that, by the end of the century, half Britain's agricultural land could be in the hands of finance institutions; this could exacerbate problems since holdings might, in such circumstances, be treated purely speculatively rather than as integral elements in the total rural economy and landscape (9).

The dictates of modern agriculture clearly require that the farm should be treated as a business, and indeed, much of the countryside is more realistically viewed as an open factory. Faced, therefore, with the object of profit maximisation or even of economic survival within a highly competitive context, it is hardly surprising that the farmer will feel little obligation to cater for the needs of conservationist and recreationalist planning needs, unless society is prepared to compensate him to do so.

Similar features characterise the forest industry, although national forest policy has perhaps been rather less insensitive to the landscape, particularly in recent years. Over the last three centuries in particular, Britain's woodlands have been allowed to decline, and this lack of a native forest reserve has had two major impacts on the nature of the modern industry. First, since no woodland capital has existed to defray the costs of new plantations, all the capital and operating costs have had to be recouped from current yields. As a consequence, Britain has never been able to compete efficiently against countries which still possess extensive areas of forest, such as Sweden, Canada and the Soviet Union, and so we must continue to import some 90 per cent of timber requirements. Second, in order to achieve a rapid return on investment, there has been an overwhelming tendency to plant fast-growing exotic coniferous species. Economic considerations result in new plantations which are often very extensive and are thus alien both in terms of scale and species composition.

With regard to a non-renewable resource, such as a petroleum reserve, rather different considerations prevail. The limiting factor is now not the maximum annual level of yield which can be sustained in

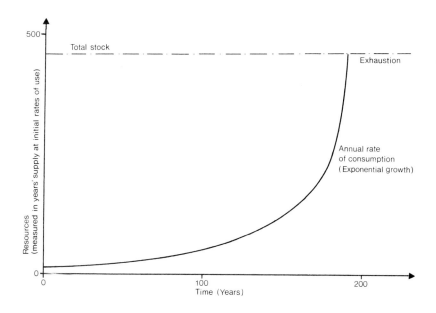

Figure 6.4 The limits to exploitation of a non-renewable resource. (Adapted from Lecomber (6).)

perpetuity — although the rate of extraction may affect the ultimately recoverable yield — but rather the total stock which is available for depletion. Once again, the rise in demand has tended to be exponential — world demand for oil, for instance, has been increasing at seven per cent per annum, resulting in a doubling of demand roughly every ten years — and this results in a surprisingly rapid exhaustion of the stock in the later stages (Figure 6.4), so that upward revisions of reserve estimates serve to delay the date of exhaustion only slightly.

In relation to North Sea oil and gas production, the UK Government has sought to influence extraction by the rate at which production licences are auctioned, by taxation measures and by fuel pricing policies. Planners have also exerted some control through their restrictions on the number of consents granted for related onshore development; the effectiveness of this type of control is clearly much greater in respect of land-based resources, such as minerals.

The Administrative Basis for Environmental Action

Town Planning in Britain – some general comments

At first sight, British town and country planning legislature seems to serve as a poor basis for ecological management, for it is almost exclusively restricted to the forward planning and site control of a very narrowly defined range of urban and infrastructural development (*s. 290(1) Town and Country Planning Act, 1971*): planners may apparently exercise only a feeble influence over land use changes in the countryside, or over industrial impacts which are not specifically in breach of a condition of planning consent.

In the immediate post-war era, when the first comprehensive Town and Country Planning Act was being framed, there was an assumption that agriculture and forestry would automatically enhance the countryside if allowed to prosper and consequently they were almost wholly exempt from planning control. Similarly, the conservation of wildlife and the control of pollution were perceived as essentially local topics to be attended to by specialist agencies, hence, no comprehensive or strategic overview was deemed necessary.

Admittedly, planners may refuse permission for, or impose restrictive conditions upon, development in the interests of amenity. However, the definition of amenity is most imprecise and may often be equated with the rather superficial concept of 'prettiness' (10). Notwithstanding this, one Statutory Instrument refers to 'detriment of amenity by reason of noise, vibration, smell, fumes, smoke, soot, ash, dust or grit', and so there may be scope for a more rigorous application of the term within environmental planning (11).

There is now a growing acceptance that planners should start to play a more central role in environmental management; and, in this capacity they already possess two distinct advantages. First, because of the uniquely synoptic nature of their profession they are in an excellent position to help coordinate the strategies of other environmental agencies. Second, many authorities concerned with the environment tend to exert a rather *post hoc* control over adverse impacts whereas planning, being essentially concerned with future land use, can seek to prevent mistakes happening in the first place.

Town planning legislature has partially responded to this changing demand. In 1965 the Planning Advisory Group recommended that the old-style development plans should be replaced by 'structure' plans, covering strategic matters, and 'local' plans, establishing more detailed planning policies (12). These principles were embodied in the *Town and Country Planning Act of 1968* (1969 in Scotland), and were consolidated with previous legislation in the 1971 Act (1972 in Scotland), which now forms the basis of most statutory planning. Following local government reorganistion, structure plans have been prepared by the County Councils (Regional or Islands Councils in Scotland) and Metropolitan Districts and local plans (which comprise district, action area and subject plans) by the Districts or Counties.

Although this new system has not significantly widened the formal planning powers in regard to ecological matters, there is greater opportunity for planners to participate in and, indeed, to co-ordinate the activities of agencies more directly concerned with the natural environment. Thus, structure plans must formulate '... the local planning authority's policy and general proposals in respect of the development and other use of land in that area (including measures for the improvement of the physical environment...)' (13), a proviso which enables planners to prepare broad ranging reviews of environmental issues and to consult with other agencies, both public and private, in the establishment of regional priorities for environmental improvement. Similarly, the greater variety of local plans enables more flexibility in resource management. Thus, for instance, *district plans* may co-ordinate measures for the protection of farming in the urban fringe, *action area plans* may provide a framework for the serving of discontinuance orders on noxious industries in inner urban areas, whilst *subject plans* may permit thematic treatment of countryside problems and their management. Very significantly, structure plans must cover the whole country and so, in theory at least, there is to be no more *white land* (which can best be described as areas left aside on development plans, and where very negative and restrictive policies apply by default), and planning policies must be established for all areas. Clearly, if such programmes are to be at all effective planners should be involved in a corporate approach both within the local authority and with external agencies; the aim should be an overall framework in which natural, artificial and human resources are managed in a long-term perspective.

Common Law in Environmental Management

Generally speaking, the right of every citizen to go about his business and to enjoy his property without being subjected to 'nuisance', is recognised under Common Law (14). In particular, the concept of 'public nuisance' might be considered germane to instances of environmental neglect or pollution. A public nuisance may be defined as an act or omission which materially affects the reasonable comfort and convenience of the life of a class of Her Majesty's subjects, thus requiring that several people in a given area must be affected, but that the disamenity must not be so widespread as to make the identification of a 'class' impracticable. However, the character of the locality must be taken into account in determining whether an action constitutes a nuisance, since everyone must be prepared to submit to a reasonable amount of discomfort consequent on the circumstances of the area in which he lives and the trades and activities carried on around him.

In addition to the concept of nuisance, the rule of *Rylands v Fletcher* imposes strict liability upon the occupier of land who keeps upon it anything likely to do damage if it is accidentally released (the original case concerned an impoundment of water which escaped down disused mine shafts). In such circumstances there need be no negligence and the plaintiff need have no connection with the land, although it must be shown that there was 'an accumulation and an escape' and that the escaping material was potentially dangerous. Finally, there are certain environmentally detrimental incidents where 'negligence', or the breach of a legal duty of care owed by one person to another, can be proved. However, this — unlike the examples given above — does require the plaintiff to demonstrate that the defendant was under a legal duty to take care. Where actions lie *in tort*, they may have an advantage over statutory measures, in that the courts may award damages or may make a discretionary order requiring the defendant to refrain from his activities or even to carry out some positive restorative action. However, such remedies are of limited use to the planner for two reasons. First, they could only be used in the event of fairly isolated cases of pollution or, perhaps, damage to wildlife, and their use would depend largely on the initiative of individuals. Secondly, Common Law actions have rarely been taken in the case of environmental matters, probably because the litigant, if unsuccessful, would have to pay substantial costs, and few would be prepared to do this, especially when acting 'in the common good'.

Statute Law in Environmental Management

Pollution control

Where a clear case of nuisance due to pollution exists, the public may reasonably expect the appropriate authority to utilise its powers of Statute Law to curb it. For example, where a local authority is satisfied that a statutory nuisance exists, it must serve an abatement notice on the person by whose act, default or sufferance it arises; if this person cannot be traced, the local authority itself may take steps to abate the nuisance and prevent a recurrence (*s. 93, Public Health Act 1936*).

The broad fields of pollution control have grown up, understandably, in response to public health hazards, in particular those which were associated with the dramatic population rise and new manufacturing processes during the Industrial Revolution (15). The passage of a number of pioneering enactments — for instance, the *Public Health Acts 1848 and 1875*, the *Alkali, etc., Act 1863*, the *Labouring Classes Lodging Houses Act 1890* — established the joint basis for the Public Health and Town Planning movements. Whilst only minor planning statutes were introduced in the early part of the 20th century, however, the passage of the *Public Health Act 1936*, provided a quite comprehensive coverage of matters of general hygiene, such as drains, sewers and scavenging, and of personal health. This continuing emphasis on the reduction of disease and epidemics has, historically, tended to relegate the broader issues of amenity and environmental quality to a very secondary role.

The main body of legislation concerning pollution control has conventionally been based upon two premises. Firstly, whilst it is recognised that a certain amount of pollution is inevitable, it is accepted that such discharges as do occur should be subject to specific conditions of consent laid down by a competent control authority, and secondly, it is presumed that, in the interests of equity, both the undertaking and the members of the public affected by it should be given equal consideraton. Consequently, conditions of consent, whilst not being unreasonably restricting, should seek to ensure that the potential damage is minimised; hence, they are often determined by reference to the criterion of 'best practicable means', enabling standards to be made more stringent over the years as technological improvements permit effluents to be treated with greater economy and efficiency.

The statutory basis for pollution abatement is thus gradually becoming increasingly comprehensive and effective. In particular, the introduction of the *Control of Pollution Act* (1974) consolidated the individual enactments which had grown piecemeal over the years, and has provided not only for the more co-ordinated operation of the various control authorities, but has afforded a basis for the collection, publication and public discussion of a great deal of relevant information which would hitherto have been confidential. The Act does have certain drawbacks — the comparatively minor penalties which most infringements carry, and the slow rate of implementation of many sections — but nevertheless its virtues greatly outweigh its deficiencies (16).

Although the *Control of Pollution Act* has made few powers directly available to planning authorities, opportunities exist for planners to conduct and publish surveys of pollution levels in their areas, to establish a priority system for the eradication of regional pollution problems, and to assist in the formulation of a common programme of action for the several agencies which do have a statutory remit for controlling environmental pollution. Planners have been urged to consider 'the need to combat and prevent pollution, even though responsibility and specific action in this field will often be for bodies other than the local planning authority' (17). Similarly, all public agencies are required to conduct their activities with due regard to the 'natural beauty and amenity of the countryside' (18).

The various authorities concerned with pollution control may exercise a variety of interventions at different stages of the manufacturing process, namely, the initial location of development, the actual process of manufacture, the nature of the composition or construction of the product, the method of waste treatment, and the place, rate and method of disposal of wastes ('end-of-pipe' control). Of these, the planning authority exercises a primarily locational control over new developments and waste disposal activities, and may impose appropriate conditions on the grounds of 'amenity'. This is not a particularly sensitive method for, once a planning decision has been made, it can only be altered by serving a discontinuance order, revoking permission or modifying conditions of consent, procedures which all involve confirmation by the Secretary of State and the payment of compensation. Pollution standards may be laid down as planning conditions, although these could not subsequently be varied, and might prove extremely difficult to enforce. It is therefore at the strategic level that the planner is most likely to play an important part in pollution control, both by influencing the broad pattern of spatial relationships between polluting activities and the environment, and by surveying regional environmental quality and establishing clear priorities for its improvement (19).

Pollution of freshwater

The pollution of freshwater is of concern to local authorities partly because of their responsibility in the provision of a potable water supply, and for the safe disposal of sewage to inland waters (*cf Public Health Acts 1936 and 1961; Public Health (Drainage and Trade Premises) Act 1937; Sewerage (Scotland) Act 1968*). Regional Water Authorities (in England and Wales) also have powers, under the *Water Resources Act of 1963, 1968 and 1971*, to control potentially polluting discharges into underground strata and to pass bye-laws regarding the pollution of water intended for domestic use. (In Scotland, parallel powers are possessed by Regional and Island Councils, and these are supported by the technical and advisory services of the River Purification Boards). The general protection of water quality is now covered by Part II of *the Control of Pollution Act 1974*, which has achieved a more integrated approach to the definition of waters subject to control and the nature of the pollutants. Broadly speaking, it makes it an offence to introduce, without permission, any effluent which might have a deleterious effect not just on public health but, very significantly, on flora and fauna also, into any inland watercourse, 'specified' underground water, or 'controlled' or 'restricted' estuaries and inshore waters. With a few minor exceptions, an operator wishing to discharge wastes is required to obtain a licence from the relevant authority, which, if granted, may be made subject to conditions regarding the nature and frequency of the discharge, the keeping of records and provision of monitoring facilities, and measures to prevent the discharges from entering specified underground waters. Farm wastes are less easily controlled; apart from the problems of locating sources, discharges are not subject to control if they are in accordance with 'good agricultural practice'.

By the early 1970's, responsibilities for water supply and drainage in England and Wales were divided between some 29 river authorities, 160 water undertakings, and 1,200 sewage disposal authorities, and this rather haphazard administration inevitably led to inefficient duplication of effort and to the frequent intervention of the DoE in order to resolve conflicts (20). Accordingly, the *Water Act, 1973*, established nine all-purpose Regional Water Authorities for England, with boundaries based primarily on river catchment areas, and the Welsh National Water Development Authority. These agencies are presently responsible for water supply and conservation, river pollution control and river management, sewerage and sewage disposal, and fisheries and land drainage.

In addition to their purely hydrological duties, the RWA's may, under the permissive powers of the 1973 Act, make provision for recreation and conservation interests, and they receive advice and assistance on these matters from the Water Space Amenity Commission and, to a lesser degree, the British Waterways Board. Thus, whilst attitudes to public access in water catchment areas, or to the large scale recreational use of reservoirs, have traditionally been highly restrictive, the more widespread installation of secondary treatment plant in recent years has greatly enhanced the prospect for the multiple use of inland waters.

Marine pollution

The administrative arrangements for marine pollution tend to be relatively fragmented: the Minister for Agriculture has a general responsibility for preventing the pollution of fishing grounds, the Trade Secretary is responsible for the control of oil and chemical pollution from sea-going vessels and for remedial action at sea, and the Energy Secretary has a concern for offshore oil and gas operations. The Secretary of State for the Environment (or the Secretary of State for Wales, or Scotland) attends to the protection of the shoreline from pollution, and requires coastal County Councils in England and Wales (Regional and Islands Councils in Scotland) to draw up contingency plans for oil pollution in consultation with the District Councils, who must do most of the cleaning up. A variety of legislation exists in respect of marine pollution but, unfortunately, it is largely restricted to ships registered in certain countries or to incidents which occur in territorial waters.

Atmospheric pollution

The campaign against atmospheric pollution has a long history in Britain, and indeed the *Alkali Act of 1863* is considered to have been the first effective national pollution control act in the world. This Act, together with its numerous successors, has accorded extensive powers to the Alkali Inspectorate, in particular enabling it to require operators to adopt the best practicable means of preventing the escape of noxious gases into the atmosphere. Complementary to these powers have been the provisions of the *Clean Air Acts 1956 and 1968*, which have enabled local authorities to designate Smoke Control Areas, and have empowered them to specify chimney stack heights in proposed industrial developments. *The Health and Safety at Work Act 1974* has now replaced the greater part of the Alkali Acts, and has established a Health and Safety Executive which incorporates the Alkali Inspectorate.

Radioactive materials

Control over pollution by radioactive materials, under the *Radioactive Substances Acts 1948 and 1960* and the *Radiological Protection Act 1970*, is primarily concerned with the protection of those who come into contact with them in the course of their work, rather than with the safety of the community at large. The day to day responsibility for nuclear installations lies with the Health and Safety Commission, which reports to the Secretary of State for Energy, whilst the disposal of nuclear wastes is the joint responsibility of the Environment and Fisheries Ministers. Local authorities must accept and deal with radioactive wastes sent to their tips, although the Secretary of State will consult with the authority before granting authorisation.

Waste disposal

The negligent pollution of land has decreased since tipping, by both public and private agencies, is now controlled in accordance with planning conditions. The legacy of derelict land which was bequeathed to many parts of the country has been the subject of several pieces of legislation, as a result of which generous grants have been made available for the purposes of rehabilitation. The most significant reclamatory programmes have been those accomplished under the provisions of the Local Employment Acts and the Scottish and Welsh Development Agency Acts.

Nowadays, perhaps the major land pollution problem arises from refuse disposal; and most household waste, and much of that arising from trade and commercial premises, come under the direct control of refuse collection authorities. In addition to the general scavenging requirements of the *Public Health Act 1936*, Councils must also provide places where residents may conveniently and free of charge, dispose of refuse (*Civic Amenities Act 1967, Refuse Disposal (Amenity) Act 1978*). Under Part I of the *Control of Pollution Act 1974*, disposal authorities must carry out a survey of the wastes requiring disposal in their areas, and prepare plans to ensure that satisfactory facilities are made available; in addition, they are encouraged to investigate opportunities for the recycling and reclamation of refuse.

Minerals

For the local authority planner, action in respect of mineral development is severely restricted but some areas of influence do exist. Firstly, at the most strategic level, planners may influence depletion policy, in the light of the relative social and economic need to exploit a limited resource at a given point in time, or to conserve it for the benefit of future generations. Secondly, they may seek to prevent mining activity altogether in areas of exceptional beauty, such as the 'National Heritage Areas' proposed by certain members of the National Parks Policies Review Committee (21). Thirdly, the precise scale and siting of new workings can be influenced by planners, as can the impact of operations and the proposed restoration measures.

The Stevens Committee, which investigated the need for changed methods of control over new mineral working, identified a number of problems peculiar to extractive operations (22). These included the destruction of land during development, the unusually high degree of understanding of the processes involved which planners are expected to acquire, the transitional nature of the operations and the consequent need to consider an after-use, and the extent to which operations could increase in scale over time or could become unacceptable in the light of changing social values. The Report suggested that control of minerals should be subject to a 'special regime', which would be implemented on both strategic and local scales (the local scale in this case being the *County* planning authority). Matters appropriate to the structure plan included general strategies for mineral workings and their after use, the protection of reserves from the encroachment of development which might sterilise their future use, and the criteria against which minerals applications should be assessed. At the local level, it was recommended that a special proforma for application should be introduced, and that it should be possible to monitor operations and regularly review the conditions of planning consent.

The Department of the Environment has recently responded to these recommendations and, whilst not accepting the idea of a separate development control procedure, it has endorsed a number of important points (23). In particular, it has provided that the re-working of waste tips should require planning permission, that a maximum life should be placed on a mineral permission, that new procedures should be established to ascertain when workings have effectively ceased in order that after-use conditions may be implemented, and that more rigorous restoration conditions should apply. One of the more controversial proposals is that investigatory operations should no longer require permission, and this has been opposed by amenity groups on the grounds that the application to prospect provided them with an early warning system when operators were interested in exploiting a particular area. Most of these provisions are included in the *Town and Country Planning (Minerals) Bill,* currently before Parliament at the time of writing.

Wildlife

The need to conserve the countryside has long been recognised by central government. The *National Parks and Access to the Countryside Act 1949* led to the establishment of National Parks and Areas of Outstanding Natural Beauty, as well as providing for a Nature Conservancy with the power to acquire and manage National Nature

Plate 15. An amenity screen at Nuneaton Quarry, Hartshill, illustrates the scope for reducing the impact of mineral extraction.

	Appraisal of ecological resources	Ecological engineering and design	Ecological management
Wildlife resource management activities	Environmental Impact Statements Evaluation studies Land Use Species richness Habitat deversity	Balancing lakes Grasslands Grid road woodlands Other woodlands Hedgerows Sewerage treatment and disposal	Biocides Pollution Silviculture Grazing/mowing Public pressure Recreational lakes
Stages of new town development	CONCEPTION →	DESIGNATION → IMPLEMENTATION →	CITY
Documentary activities	Historical ecology (Complements archaeological activities) -pollen-snails-geology-soils-vegetation of archaeological sites	Detailed investigation of ecosystems and habitats (Biological components/ Quantitive studies/ Site histories) Woodlands-Grasslands- Rivers-Gravel pits- Clay pits-Spoil heaps- Ponds-Road verges- Hedgerows-Railway line- Arable land-Canals	Rescue Transplanting Recording Museums

Monitoring ecological effects of:

Urbanisation
Recreation
Public pressure
Management
Colonisation of new habitats

Table 6.1 Wildlife conservation and the local authority.

Reserves. These measures have ensured that we have retained the best of our countryside, but they have afforded little protection to the rest of it: wildlife reserves, for instance, account for less than one per cent of our land area, and the bulk of our wildlife thus lives on ordinary farmland where little or no reprieve is given from the ravages of mechanisation.

Local planning authorities may assist in the development of conservation strategies in a number of ways (Table 6.1). They may perform an executive role, by managing designated sites or in creating new ones in the course of landscaping or reclamation schemes; they may assist in environmental education by providing interpretive facilities and nature trails and by the dissemination of information and advice; they may act in a consultative capacity, liaising between developers and conservationists; and they may play a protectionist role, seeking to channel pressures away from zones of ecological importance, or to control development to ensure the continued existence of site features worthy of preservation. In addition, the local authority may consider it their duty to assist in ensuring the public's observation of the *Conservation of Wild Creatures and Wild Plants Act 1975*, which provides protection for wildlife generally, as well as making special provision for the rarest of our flora and fauna.

The traditional approach to wildlife conservation has been to purchase and manage sites of outstanding value, and local authorities are enabled to designate Local Nature Reserves (*National Parks, etc., Act 1949*) and may also include nature conservation objectives in management plans for Country Parks. In many such sites, education will probably play an important role and, consequently, the major management factor will be to strike a balance between the requirements of the resource in terms of its sensitivity to trampling and disturbance and its accessibility to the public. The most common method of channelling visitor activity whilst still achieving educational objectives is by introducing nature trails and these vary in their degree of formality according to the level of control desired. Formal trails may be laid out to provide a rigidly guided tour between points possessing some specific interest, whereas semi-formal trails provide a set route but only treat habitat and fauna in broad terms, whilst informal trails simply furnish information which allows the user to walk generally within the area, seeking out the features of most interest to him (24).

For too long, however, there has been a tendency to treat nature conservation as if it concerned only the protection of a few designated sites and species. It has only been in recent years that there has been a realisation that conservation is vitally concerned with the dynamic processes of change and adaptation, and that the success of a network of isolated reserves depends in a fundamental manner upon the diversity and balance of the countryside as a whole. Perhaps the most important role which planners can currently play in nature conservation, therefore, is to ensure that the needs and aspirations of the various conservation bodies are taken into account, and that the incremental and cumulative effects of day-to-day developments do not erode insidiously the general quality of the environment (25). The consultative role of the planning authorities in deriving broad strategies together with the detailed scrutiny which can be afforded to routine projects constitute the planner's main powers for conservation, and thus a recent Circular exhorts local authorities 'to take full account of nature conservation factors both in formulating structure and local plans, and in the consideration of individual planning applications' (26).

In certain instances it can be that a formal arrangement is desirable and the use of management agreements between LPA and owner or tenant whereby the latter adheres to certain conservation practices in return for compensation and possibly warden services from the former, has been widely advocated (27). Often, though, the success of these is more illusory than real, and Lord Porchester has recently drawn attention to the formidable obstacles to their successful operation, including the inability to bind successors in title, the lack of statutory back-up powers, the absence of a prescribed basis for compensation, their reliance on restrictive rather than positive covenants and the difficulty of persuading a farmer to change what may already have become firm commitments (28). The operation of a 'notification system' is a desirable adjunct to management agreements so that planners are given advance warning of adverse developments. In the Exmoor National Park, for example, farmers within a specially defined 'critical amenity area', when applying for improvement grants, are advised by the Ministry of Agriculture that they should approach the National Park Authority with a view to reaching a compromise solution, if one is possible.

At the time of writing, a Wildlife and Countryside Bill which seems certain to give legal substance to management agreements, and to ensure that these run with the land (rather than with tenant or owner) is before Parliament. Additionally, in National Parks, local planning authorities may make grants or loans for conservation work. However, the important proposals for Moorland Conservation Orders have failed to survive a change of government, although lobbying may lead to their inclusion as last resort back-up powers.

Under the *Nature Conservancy Council Act, 1973*, the old Nature Conservancy was effectively split into an advisory body (the NCC) and a research branch, the Institute of Terrestrial Ecology. Nowadays, in addition to their traditional roles of commissioning research, managing reserves and disseminating information, the NCC also advises

Ministers on national nature conservation policies, whilst their regional officers must give advice to planning authorities regarding the ecological effects of development proposals. However, the Council is hard pressed to give adequate attention to any but the most urgent cases, and it is therefore often more expedient for planners to seek the advice and detailed local knowledge of voluntary organisations, such as the County Trusts and Royal Society for the Protection of Birds. At present it appears that this potential is often under-utilised through an absence of standardised consultation procedures, a lack of ecological knowledge on the part of the planners and a shortage of suitable information amongst the conservationists. A number of initiatives could be taken by planners in order to rectify this, for instance, the automatic notification of Trusts with adequate information regarding certain classes of proposals, involvement of Trusts at an early stage in the preparation of plans and the co-opting of Trust members onto planning working parties. For their part, Trusts could endeavour to develop comprehensive information and record systems on local sites of ecological significance (29).

Agriculture

The rural estate management lobby has always been vocal and politically powerful; this has largely accounted for its success in resisting calls for the extension of some kind of planning control over farming and forestry. However, much of this opposition can be seen to rest on a false premise, for the system of Exchequer grants and subsidies and a variety of government advisory services on land management already perform an extensive regulatory role. Thus, resource mangers' claims that they do not wish planners to 'interfere' in their activities are rather hollow, and it is reasonable to suppose that, in the future, the work of the various countryside agencies will become more formally co-ordinated both in spirit and purpose.

So far, the primary involvement of planners in regard to agriculture has been to protect land from low density urban sprawl, and for many years it has been official policy to ensure that, wherever possible, agricultural land of a higher quality is not taken for development where land of a lower quality can reasonably be used. More stringent guidelines have recently come into force, in particular requiring planners to consult with the Ministry of Agriculture regarding development of greenfield sites over ten acres (30). Local Planning Authorities may also exercise control over farm buildings where the ground area exceeds 465 sq metres, the height exceeds 12 m, or it is within 25 m of a classified road. This ensures that most intensive livestock units — the most controversial on amenity and pollution grounds — can be subject to control, and a number of planning authorities now exercise a comprehensive policy with regard to such development (31).

In addition to these roles, the Agriculture Economic Development Committee has recommended that planners should encourage the greater re-use and restoration of derelict and under-utilised land, seek to minimise the fragmentation of farms affected by development proposals, promote the improvement of rural facilities for agricultural communities, and designate 'buffer zones' in the urban fringe as a means of reducing indiscriminate recreational usage of farmland (32). The Strutt Committee (33) has further recommended that planners should ensure, by the imposition of appropriate planning conditions, that farmland earmarked for development but not yet required should be acceptably farmed by the developer in the interim; that the present 'ten acre' threshold for consultation with the Ministry of Agriculture should be reduced to five acres; and that call-in procedures should be strengthened, so that the DoE, at the Ministry of Agriculture's request, should call-in any case where there is an unresolved agricultural objection.

Despite these various powers possessed by planners, the principal agent of public intervention is, of course, the Ministry of Agriculture, who, by the sytem of grants (notably the Farm Amalgamation Scheme and the Farm and Horticulture Development Scheme) and subsidies (for instance, assistance under the EEC 'Less Favoured Areas' Directive) which they administer, influence the broad structure of British farming. This influence is extended to the local level by the field officers of their Agricultural Development and Advisory Service (since 1970, ADAS has been able to offer farmers limited advice on conservation matters). Thus, there are, it has been contended, two types of planner concerned with farming: the agricultural planner (of MAFF or ADAS), to whom farming regions are units of production within a single national farm, and the town and country planner, to whom farming regions are landscape tracts with particular socio-economic and settment features to be taken into account (34). However, the two were brought slightly closer together in the summer of 1980 when regulations were introduced to require farmers to obtain approval for grant-aided improvements in National Parks and AONBs (from the Local Planning Authority) and in SSSIs (from the Nature Conservancy Council). Any unresolved objection must now be determined at Ministerial level.

Where wildlife or amenity considerations are of a particular interest on a farm, two modes of action are possible: either a purely voluntary approach to reconcile the competing interests, or a more formal attempt to strike a management agreement between farmer and local planning authority. Many local authorities also own farms themselves, and thus it may sometimes be practicable to run these partly as demonstration units to show the scope for compromise between profitability and conservation. Suggestions have also been made that the Ministry of Agriculture should consider making improvement grants subject to amenity conditions, as occurs in some European countries.

The scope for voluntary initiative has greatly broadened since the 1969 'Silsoe' Conference on farming and wildlife, and the subsequent establishment of a Farming and Wildlife Advisory Group (FWAG). Local naturalists' trusts have furthered the work of FWAG, for instance by bringing the various advisory publications and codes of practice on farming operations published by MAFF and conservation organisations to the attention of farmers, whilst in West Wales the local Trust has succeeded in establishing Farm Nature Reserves (35). Practical assistance in the establishment and maintenance of wildlife habitats on farms may often be provided by voluntary organisations, notably the British Trust for Conservation Volunteers. Clearly, this is a field in which local planning authorities may wish to assume the role of mediator.

Forestry

The second major land user of the countryside, and one whose ecological impact has in many respects been comparable to agriculture, is the forest industry. Under the *Forestry Acts* of 1967 and 1974 the Forestry Commission has been required to give greater attention to the aesthetic and recreational role of public forests. Moreover, the private foresters, who receive planting and management grants under the Basis III Dedication Scheme set out in these Acts, are bound by its provisions to give attention to conservation, recreation and effective integration with agriculture.

In Scotland, the Forestry Commission has invited planners to co-operate with them in devising regional forest strategies (36). As a first step, it is advocated that plans should ascertain the present status of forestry within the region, paying particular regard to the extent and nature of existing woodlands and hedgerow trees, levels of timber production, related employment demand, and the role of woodlands in the environment. Secondly, it is proposed that elements favouring increased forest development, such as topographic and economic criteria, and the potential for integration with other forms of land use, should be discussed alongside a *précis* of the salient features of the appropriate Conservancy Recreation Plan. Finally, it is intended that areas should be identified where there would be a strong presumption against further afforestation on the basis of distance and housing constraints, or of adverse impacts upon environmental, farming and recreational interests.

In addition to co-operation over the strategic development of timber reserves, planners are also concerned, in part, with local forestry matters. Felling Licences are generally required for the harvesting of mature trees (except in the case of Dedicated Woodlands), and applications are referred to the local planning authority for observations before consent (which may be made subject to conditions) is granted or refused. However, it is permissible to fell roughly a dozen trees per quarter without sanction, and, where this quota is exceeded illegally, the threat of a sizeable fine is remote. Planning authorities may also impose Tree Preservaion Orders, although since these cannot be used to enforce good silvicultural practice, they should only be used as a last resort.

Despite the various measures available to planners, the tree cover of our lowlands has been depleted to such a dangerously low level and has acquired such a preponderance of ageing trees, that it is in danger of extinction. A concerted effort aimed at the regeneration of small woodland and hedgerow trees is belatedly underway. This campaign is based on direct action by public agencies, and on the provision of information to private individuals (37). Certain grants are available from the Ministry of Agriculture for the establishment of shelter-belts and small woodlands on farms, whilst others may be obtainable for amenity planting from the Countryside Commission or from local authorities, and, where enhancement of wildlife interest is the principal aim, from the Nature Conservancy Council. Planning authorities can, of course, undertake tree planting schemes of their own, and the need to direct funds towards the renewal of environmentally depressed urban areas should not be allowed to divert attention from the equally urgent task of the re-creation of rural landscapes.

Public Participation

Environmental law is, hopefully, not simply implemented by remote bureaucrats on the basis of professionally preferred strategies. Planners are instead subject to public accountability by elected councillors and thus all planing decisions are, ultimately, political, rather than technical.

The councillor system has not, however, always kept as close a check on over-zealous local government activity as it might. The difficulties faced by part-time elected representatives of keeping in touch with the views of their ward members on a wide range of issues has, on occasion, led communities to feel that they were being inadequately represented. Since the late 1960's in particular, and thus coinciding with the resurgence of interest in environmental matters generally, direct action has been the order of the day over contentious local issues. Consequently, safeguards have been built into planning and other environmental laws.

Plate 16. Our lowland woodland cover is becoming seriously depleted – imme-diate and far-reaching measures are required to ensure its regeneration.

Since 1947 it has been necessary for all local planning authorities to maintain a register of applications for planning permission and of their decisions which must be open to public inspection. In respect of applications for certain especially unneighbourly types of development, or for development within conservation areas, publicity must be given to the application by means of press advertisements and site notices, and representations may be made by objectors to the LPA within 21 days. Very controversial applications may be called-in by the Secretary of State for determination, and become the subject of a quasi-judicial Planning Inquiry (or Public Inquiry); Inquiries may also be called over other matters, notably proposed Local Plans or the refusal of planning permission. Objectors to the scheme must be notified of the Inquiry, at which they may be called upon to give evidence. So complex have Inquiries become in recent years that now a 'pre-inquiry' is often held to sift out the major issues for more detailed investigation at the main event. Under the *Control of Pollution Act*, registers of disposal licences must be kept open for public inspection and the public must be able to obtain copies of entries. With regard to water pollution, the relevant authority must publish a newspaper advertisement and allow a period of six weeks for representations.

Such modes of consultation, however, provide the concerned environmentalist with only a sporadic and reactive input into the development process; the 1968 and 1971 *Town and Country Planning Acts*, therefore, provide that an opportunity to contribute to forward planning should be given. Before formally preparing a structure plan, the County planning authority must give adequate publicity in their area to their 'Report of Survey' on which the structure plan is to be based, and also of all the matters which they propose to include in this plan. Any person or organisation who may be expected to have a particular interest in these matters is entitled to an adequate opportunity to make representations to the local planning authority. Once prepared, the LPA must submit the plan to the Secretary of State and provide copies for public inspection, with the public being given opportunity to make written objections to the Secretary of State. Before determining whether or not to approve the plan, the Secretary of State will appoint an Inspector to hold an 'examination in public', at which 'key issues' selected by him will be considered by invited interested parties. With regard to the adoption of a local plan, the LPA must give similar publicity to its proposals, and invite representations; once prepared, objections must be made to the LPA itself. There is no examination in public, but a private hearing or public local inquiry may be held into any objection. Under the *Control of Pollution Act*, similar provisions for representation exist regarding the making of a Waste Disposal Plan or a Noise Abatement Zone.

Clearly, there is considerable scope within the framework of environmental law for amenity groups to make their voices heard, but there are still serious limitations. Often, the agencies of government, especially the statutory undertakers, are exempt from the need to apply for planning permission, and thus the LPA is reduced to the status of a consultee. In the USA, federal agencies must have major projects validated by publicly scrutinised environmental impact statements, which clearly provides for a more democratic solution. In the UK, the call-in procedure of major applications by the Secretary of State is also arbitrary and so there is no guarantee that a Public Inquiry will be held in any particular case: there was, for instance, a very marked reluctance to call-in the application for the proposed reprocessing plant at Windscale. Moreover, the outcome of the Inquiry — the Inspector's recommendation — is not binding, but is made subject to political approval: thus, objectors have been known to 'win' their case, only to have the recommendation overturned by the Secretary of State, as for instance occurred over the proposal to extend Edinburgh airport.

The objectors' case at Public Inquiries has further been made more difficult by the unwillingness of developers — even where they are government departments or nationalised industries — to disclose information: until recently, for instance, the Department of the Environment refused to discuss the basis for their traffic forecasts in respect of new roads. Once more, the situation in the USA is more satisfactory, where the Freedom of Information Act makes information more readily disclosable. The opposition to major proposals has been conducted by voluntary organisations, who must themselves meet the costs of presenting their case. Gone are the days when an amateurish presentation would suffice, and objectors now find themselves having to engage very costly legal counsel. Despite strong calls to provide assistance to objectors, the government has remained impassive.

Yet perhaps there is an even more serious strain being imposed on the Public Inquiry system. The Inquiry can cover only strictly material planning issues yet it is the only effective forum at which the public may voice their opinions about a wide range of policies which affect their future. The 1980 inquiry into the dumping of nuclear waste in the Dumfries and Galloway Region, for instance, was called only to deal with the rather minor matter of drilling test boreholes, but objectors to such proposals would clearly wish to raise far broader philosophical and ideological issues about the prospect of a plutonium based society. It is questionable whether the Public Inquiry system can withstand the strain which is being placed on it, and planners have certainly been taken aback at their role as reluctant heroes. In this sense, the planning profession has had greatness thrust upon it, and it must recognise that its duty lies not only in the day-to-day scrutiny of development, but also in opening up to public discussion the kind of future in which we wish to live.

NOTES

(1)   K. E. Boulding (1966) 'The economics of the coming spaceship earth' in H. Jarrett (Ed.) *Environmental Quality in a Growing Economy*, Baltimore, John Hopkins University Press.

(2)   See *The Planner*, 62, July 1976, pp. 124-131, for a useful collection of articles on the relationship of various concepts of 'growth' and 'no growth' to planning.

(3)   Alternatives to the GNP are discussed in H. V. Hodson (1972) *The Diseconomies of Growth*, New York, Ballantine Books, esp. pp. 50-72; and in W. W. Heller (1972) 'Coming to terms with growth and the environment' in S. Schurr (Ed.) *Energy, Economic Growth and the Environment*, Baltimore, John Hopkins University Press, for Resources for the Future Inc.

(4)   G. Hardin (1968) 'The tragedy of the commons', *Science*, 162.

(5)   P. A. Victor (1972) *Economics of Pollution*, London, Macmillan.

(6)   R. Lecomber (1975) *Economic Growth Versus the Environment*, London, Macmillan.

(7)   Ministry of Agriculture, Fisheries and Food (1975) *Food from our own Resources*, CMND, 6020. London, HMSO.

(8)   Useful introduction to agricultural economics: M. J. Stabler (1975) *Agricultural Economics and Rural Land Use*, London, Macmillan; D. Metcalf (1969) *The Economics of Agriculture*, Harmondsworth, Penguin.

(9)   Centre for Agricultural Strategy (1977) *Land Ownership by Public and Semi-Public Institutions in the U.K.*, Reading, CAS.

(10) The 'amenity' concept is discussed in D. L. Smith (1974) *Amenity and Urban Planning*, London, Crosby Lockwood Staples. It is a rather superficial term, however, and a critique can be found in P. H. Selman (1976) 'Environmental conservation or countryside cosmetics', *Ecologist*, Volume 6 (9).

(11) Department of the Environment (1972) *Town and Country Planning (Use Classes) Order*, S1 1972, No 1385, London, HMSO.

(12) Planning Advisory Group (1965) *The Future of Development Plans*, London, HMSO.

(13) *Town and Country Planning Act* 1971, s 7(3).

(14) For a discussion on Common Law in relation to environmental management, see D. A. Bigham (1973) *The Law and Administration Relating to the Protection of the Environment*, London, Oyez.

(15) For further information, refer to the Annual Reports of the Royal Commission on Environmental Pollution, London, HMSO, especially the Fourth Report (1974) *Pollution Control: progress and problems*.

(16) For a critique of the Control of Pollution Act, 1974, see F. Sandbach (1977) 'Has this Act any teeth?', *Ecologist*, 7.

(17) Department of the Environment (1971) *Memorandum on Pt. 1 of the Town and Country Planning Act* 1968, Circular No. 44/71, London, HMSO.

(18) *Countryside Act*, 1968, s. 11; *Countryside (Scotland) Act*, 1967, s. 68.

(19) C. Wood (1976) *Town and Country Pollution Control*, Manchester University Press. For an example of a survey carried out by a planning department under the Control of Pollution Act, see Cheshire County Planning Department (1975) Cheshire, *A Review of Atmospheric Pollution*, Chester, Ches. County Co.

(20) For a review, see A. Gilg (1978) *Countryside Planning*, Newton Abbott, David and Charles.

(21) Lord Sandford/DoE (1974) *Report of the National Parks Policies Review Committee*, London, HMSO.
(22) Sir R. Stevens (Chairman) (1976) *Planning Control over Mineral Working*, London, HMSO.
(23) A summary is given in R. Caisley (1978) 'Minerals proposals give us the tools but not the men', *Planner*, 64.
(24) M. B. Usher (1973) *Biological Management and Conservation*, London, Chapman and Hall, pp. 258-63.
(25) P. H. Selman (1976) 'Wildlife conservation in structure plans', *Jour. Environ. Mgmt.*, 4.
(26) Department of Environment (1977) *'Nature conservation and planning'*, Circular 108/77, London, HMSO.
(27) M. Feist (1979) 'Management agreements: a valuable tool of rural planning', *Planner*, 65.
(28) Lord Porchester (1977) *A Study of Exmoor*, London, HMSO.
(29) N. J. Beynon and B. W. Wetton (1978) *Nature Conservation and Planning: the participation of the County Trusts for Nature Conservation,* Trent Papers in Planning, No. 78/1, Nottingham, Trent Polytechnic.
(30) Department of the Environment (1976) *Development involving Agricultural Land*, Circular No 75/76, London, HMSO.
(31) Humberside County Council (1979) *Intensive Livestock Subject Plan.*
(32) Agriculture Economic Development Council (1977) Agriculture in the 1980s, London, HMSO.
(33) Advisory Centre for Agriculture and Horticulture (1978) *Agriculture and the Countryside* (the Strutt Report), London, HMSO.
(34) A. Gilg *op. cit.* (20).
(35) C. Keensleyside (1977), 'Voluntary action in conservation', in Joan Davidson and R. Lloyd (Eds.) *Conservation Agriculture*, Chichester, Wiley.
(36) Scottish Development Department (1975) *Planning Advice Note 2, Forestry Guidelines*, Edinburgh, HMSO. Also, see G. M. L. Locke (1976) *The Place of Forestry in Scotland*, For Comm. Res. and Dev. Paper 113, Edinburgh.
(37) P. Hardy and R. Matthews (1977) 'Farmland tree survey of Norfolk', *Countryside Recreation Review*, 2, Countryside Commission.

7 Future Prospects

The adoption of ecological concepts has clearly been one of the most significant developments in recent planning theory. Nevertheless, the translation of these principles into practice has so far been limited, and there has perhaps been a tendency to concentrate overmuch on a few isolated topics of common concern to planning and ecology rather than to progress toward a more complete integration of the two disciplines.

The problems of defining a workable and practical interface have arisen as much from lack of agreement within each profession as they have from any conflict of objectives between ecologists and planners. Some conservationists claim that their primary task is to manage the wildlife stock only inasmuch as it represents a resource for genetic research and a biological basis for pest control. Equally strongly, others have contended that conservation acquires its essential relevance when it forms part of a broader, humanitarian ethic whereby a more harmonious relationship with nature may be established (1). In a similar fashion, the interest generated amongst planners for ecology has been of a rather technical and negative nature and only rarely has due importance been given to the fundamental control which the natural environment imposes upon purposeful planned action.

In spite of these limitations it is evident that the liaison between planning and ecology has not been without substance. According to one survey, almost 90 per cent of the respondent County Planning Officers believed that ecology had reached the stage at which it could contribute usefully and directly to the planning process within their own authorities. Most had at their disposal a considerable amount of information on the wildlife resources of their counties, whilst almost two-thirds had established further contact with scientists, universities or polytechnics (2).

Encouraging though this initial response may be, it still merely scratches the surface of the potential interchange between planning and ecology. Indeed, the *New Scientist* found that planners tended to perceive 'the environment' as but one of several considerations to be taken into account in the policy-making process, and that, in the final analysis, it was often reduced to little more than a set of factors which acted as constraints on alternative plans; thus, 'by and large, a sense of the complex environmental change resulting from developments and actions' proved to be lacking from structure plans (3).

Central to this deficiency has been the relative paucity of appropriate information, compounded by the difficulty of acquiring an understanding of complex, seasonally variable ecosystems within the time constraints of planning applications. It is therefore essential that progress continues to be made on the systematic assessment of natural resource quality, on improving our understanding of basic ecological processes (such as have been examined in this text — nutrient cycles, pollutant pathways and ecosystem regeneration), and on establishing the ecological effects of current land and energy use patterns. In addition, it is essential that the 'monitoring' requirement of the structure plan is interpreted so as to include ecological systems, perhaps using selected wildlife species as indicators of general ecological health.

The responsibility for reducing the communctions gap between ecologists and planners and for ensuring the collection of pertinent information is seen to lie very much in the hands of the growing breed of local authority ecologists (LAEs), who serve a critical role as intermediaries between the two professions. At times, the LAE's credibility has been rather uncertain, perhaps proving too pure for his local government colleagues and too practical for the scientific community. Hence, in terms of his operational remit, he often appears to be little more than a low-cost landscape architect, a pragmatist capable of finding inexpensive solutions to the maintenance and establishment of vegetational cover on problem substrates — whereas in terms of personal persuasion he would probably prefer to contribute to the derivation of planning strategies in a more fundamental manner.

One way in which we might identify areas in which this contribution might take place is by considering some examples of current practice. These examples have not been selected by virtue of their comprehensiveness or individual excellence, but rather because they serve as indicators of possible directions for further development. As we draw to the close of this text, it seems appropriate to reflect on the kind of themes which might be pursued in plans of the future.

*Example 1: Ecology and planning **stricto sensu***

Not infrequently, strictly scientific ecological considerations form an essential input into site planning. Many award-winning derelict land reclamation schemes now lie barren because inadequate advice was sought about the kind of vegetation likely to succeed in the long-term, whilst, for similar reasons, most of the trees established during the 'Plant a Tree in '73' campaign have since died. This is an unnecessary waste, for ecological advice could have been readily obtained and followed through, especially for the sites in local authority ownership.

One example of publicly owned rural land which affords opportunities for biological management is the Country Park. In 1976 the Countryside Commission placed a one year contract with the Institute of Terrestrial Ecology to consider the management of grassland and heath in country parks, with the object of determining the efficiency, in terms both of conservation and cost-effectiveness, of alternative management techniques (4). The study identified the difficulties of trying to reconcile the presence of large numbers of people with the maintenance of a countryside atmosphere. The latter could be achieved by using a variety of farming methods, such as grazing, hay-making and cutting regimes, but these could be adversely affected by visitor pressure. It might, however, be possible to identify zones where low, medium and high public use would be appropriate, and to contrive patterns of usage by the careful siting of car parks and other access points conveniently based close to features of interest. Since the variety and abundance of wildlife is broadly determined by vegetation management, it is within the scope of the manager to create within these zones a variety of habitat types which will differ both structurally and floristically.

In general terms, it was found that grazing with cattle, sheep and deer, either alone or together, is likely to be an acceptable form of management which also adds to the desired country atmosphere, whereas cutting and burning entail problems of amenity, cost and safety. In many cases, it proved possible to lease grazing land in country parks to local farmers, thereby generating much-needed income for the management budget.

Example 2: Ecology and the environment of childhood

We may be certain that environmental variety is a major determinant of the emotional and psychological make-up of children, and nowhere is this more apparent than in their need for play and adventure in the open spaces around their homes. Through their first-hand experience of wildlife and semi-natural vegetaion, they not only acquire an appreciation of their place in nature, but they become the conservationists of the future.

Marion Shoard (5) points out that the countryside is a store of things that can be turned into toys (like conkers), weapons (like bullrushes), decoration (like daisy chains), sound (like the screech made by blowing between two blades of grass) and food (like blackberries). Unlike the sterile and formal playgrounds so often laid out by local authorities, it affords innumerable places to hide, and allows children to tumble about on grass rather than on concrete. Chris Baines (6) looks to the similar role which derelict land plays in urban areas, providing an environment rich in opportunities for 'adventure, discovery and mild destruction'. Whilst acknowledging the need to reclaim dangerous and unsightly ground, it is argued that ways must be found of making sites safe without removing all scope for exploration: of re-vegetating slopes without resorting to uniform drainage, levelling and scrub clearance.

Children must learn not only of the wildlife and freedom of the countryside, but also about its activities. Many urban children now believe that eggs originate from cartons and that milk is produced in bottles. An important attempt to introduce children to the world of agriculture is being made by the City Farm Movement, a voluntary organisation which has received considerable financial support from both public and private organisations. A typical list of features on a city farm might include animal pens and chicken runs, community gardens, picnic areas, stables and hay-sheds; all these are likely to be managed substantially by local volunteers.

Perhaps the most salient advantage of 'natural' play areas is not so much that they are inherently desirable from the point of view of environmental education but, as is so often the case with ecologically-based solutions, that they are relatively inexpensive, easy to maintain and successful. Many 'planned' play areas are so obviously the product of officialdom that they have suffered repeated vandalism: proponents of the City Farm Movement, conversely, point to the insignificant levels of damage which they experience, and to the high degree of commitment to them which many 'problem' children display.

Example 3: The ecology of the living environment – 'Environment' in the West Bromwich Structure Plan

Just as our surroundings influence our behaviour during childhood, so the quality of our home environment contributes to our sense of wellbeing throughout life. The West Bromwich Structure Plan Report of Survey interprets the term 'environment' in this broad sense, and it illustrates well the way in which the planning framework can be used to tackle a number of related problems in a coordinated and concerted manner (7).

Plate 17. Through their involvement in improving the environment, children not only acquire an appreciation of their place in nature, but they become the conservationists of the future. Above, pupils of Greenhall High School, Gorebridge, at work on their prizewinning project at Mellon, Udrigle, Wester Ross.

The definition of 'environment' adopted by the planners conforms closely to our conception of human ecology. The Report of Survey states that:

"... the environment occurs within the physical framework of the urban unit. (Human actions), together with the conditions under which those actions take place, join with the functional and aesthetic quality of the buildings ... the open spaces, and the land form and natural features of the area to create an environment".

When certain aspects of this living environment are perceived as being deficient, planned intervention becomes necessary.

Planners in West Bromwich therefore decided to carry out an environmental study of the area covering as wide a range of components as possible — the preliminary investigation ranged from townscape to crime rates and the incidence of rodents. However, this was gradually narrowed down to a set of the most important factors, which were then plotted on overlays showing the incidence of critical levels of environmental deficiency. For example, local shopping, educational, health, welfare and public transport facilities were appraised on the assumption that houses over a quarter of a mile from such a facility would suffer a deficiency. Included in the resulting criteria were: housing age, condition and occupancy; road patterns and accident black spots; industrial intrusion and pollution; recreational facilities, open space and vegetation.

Once the extent of the problem had been surveyed, it became possible to advance a set of aims and objectives, and to propose planning policies appropriate to their attainment. Of the many objectives set forward, a few may be selected to illustrate the scope of the plan, for instance, 'the reduction of the detrimental effect of working establishments on the natural environment', 'the reconciliation of interests between agriculture and leisure', 'the protection of areas of natural, architectural, historic and archaeological interest' and 'the provision of all substandard dwellings with adequate internal amenities'.

Example 4: Ecological energetics of human communities

We have seen how the study of ecology has become increasingly concerned with the efficiency with which energy is utilised, since the total amount of available energy and the way in which it is utilised impose an upper limit on the size and diversity of an ecosystem. Clearly, modern urban-industrial systems are unnecessarily wasteful in their use of energy, and part of the reponsibility for correcting this state of affairs lies with planners; one observer has argued that 'only the planning profession can bring an overall approach to energy development and evaluate the implications for society as a whole' (8).

Longmore and Musgrove (9) have identified several aspects of energy consumption which are of interest to urban designers. Urban areas, as we noted earlier, possess characteristic microclimates, which influence energy use: particularly in winter, night temperatures tend to be higher in towns, and thus heating degree-days (the amount of heating required to raise daily temperatures to a desired norm, aggregated throughout the year) are about ten per cent less than surrounding rural areas, whilst the lower mean wind speeds further reduce heat loss from urban buildings. Thermal insulation will restrict heat loss, and substantially improved standards now apply to new buildings as a result of the 1974 and 1978 Amendments to the Building Regulations. However, energy-efficient buildings designed in isolation do not offer the same scope for conservation as do planned complexes. Coordinated programmes may further afford opportunities for district heating schemes, whereby heated water is carried by lagged pipes from a central source to high density residential areas. In certain cases it may be appropriate to use incineration plants as the means of heating the water, and s.21(1) of the *Control of Pollution Act* gives local authorities certain powers of heat and electricity generation in connection with the combustion of refuse (although problems are still posed by the monopoly position of the CEGB). More consideration must be given to the relaxation of zoning controls, which enforce the separation of residential and industrial areas, so that opportunities can be provided for shared use of heat and power, and so journeys-to-work can be minimised.

The scope for optimising urban form is strictly limited because of the prohibitive costs of modifying existing infrastructure to accommodate new patterns. Most district heating schemes, for instance, have been developed in Eastern Europe, where large-scale construction of new communities has taken place under strongly centralised control. However, whilst recent years have seen comprehensive redevelopment schemes fall into disfavour in Britain, the author considers that the heyday of rehabilitating obsolescent dwellings may be passing, and that the bulldozer may be due for a comeback, albeit in a more acceptable role. If this does prove to be the case, energy considerations may well become paramount in design proposals.

The criterion of energy efficiency is as relevant to regional planning as it is to urban design. Braiterman *et al* (10), examined the contribution of 'landscape resources' to the economy of several study areas in the eastern USA. The investigation set out to determine whether the protection from urban encroachment of landscape resources (for instance, farmland, aggregates, commercial forest, water supply areas, recreation facilities) close to a conurbation would result in the provision of commodities to that community with greater energy efficiency than if they were imported. It was found that, in general, the savings of energy obtained (predominantly associated with production and transportation costs) by utilising the local landscape resources were substantial in relation to the energy costs incurred by importing products from outlying regions, even where these were top-ranking production areas. Indeed, these savings generally more than offset the loss of potential gain from urban development. Such findings suggest an increasingly important role for Green Belt policies in Britain.

Example 5: Impacts in the countryside – the Mersey Marshes Study

Concerned at the expansion of the petrochemicals industry in the area between Ellesmere Port and Runcorn, Cheshire's County Planning Department sought to produce a local plan for the area; in the event, the study on which the plan was to be based bore many resemblances to an environmental impact assessment (11). The major problem from the planning point of view in this area has been the difficulty of coping with the sustained, incremental growth of the Stanlow complex, which began in the late 1920's as a distribution depot and has grown over the years into a major oil-refining and industrial complex in excess of 800ha.. Consequently, development has now reached the natural boundaries of the existing site, and the planning authorities must decide whether to release yet more land. Two nearby sites seem likely candidates to accommodate further expansion: one is readily developable, but could present considerable amenity problems for nearby settlements, the other, the Ince Banks, would entail high reclamation costs and wildlife loss (the Mersey estuary is an SSSI) but might be preferable on social grounds.

A major concern of local residents was that of increased air pollution. Action groups from the villages of Frodsham and Helsby, which stand on a ridge above the marshes, contended that topographical and climatic conditions combined to cause high concentrations of pollution in the lee of the ridge, and that chemical emissions, particularly of lead and ammonium nitrate, were having an adverse effect on health. Monitoring of pollution levels was carried out by the local authority, and the findings appeared to suggest that the residents' fears were based on perceived, rather than real, dangers. A further area of concern was that of hazard, an issue about which planners had to liaise closely with the Health and Safety Executive in order to determine optimum buffer zones between high risk sites and residential areas. On the basis of the limited information available it was suggested that two villages were exposed to a sufficiently high level of risk to make it prudent to refuse planning permission for any future building.

A second study centred on the feasibility of reclaiming the Ince Banks, in terms of costs and natural constraints, as an alternative to allowing development to continue to drift incrementally towards residential areas. This site had been suggested during the public participation stage of the County Structure Plan on the grounds that it would present fewer problems of hazard and nuisance and that it would safeguard high quality farmland elsewhere.

A major consideration in this study was the wildlife value of the banks, and an ecologist was appointed specially to conduct an investigation of these. The main feeding areas for wintering wildfowl and waders near the south bank of the estuary would remain unaffected by development; however, their roosts would be seriously encroached upon, and this led to an unfavourable report on the reclamation proposals. Hydrological considerations also seemed to militate against the development: the Port of Liverpool Authority and the Mersey Conservator objected strongly to any reclamation, arguing that it would reduce the tidal volume of the estuary and thus restrict the natural scour of the river, leading to an increase in the amount of dredging required.

Example 6: Planning the rural scene – inter-agency approaches

Planning the rural scene is often a very different affair from planning in cities. Relatively anonymous social groupings are replaced by independent-minded farmers who remain sceptical of the need to consider factors other than the balance-sheet of their enterprise. Statutory control can seldom be a viable approach in the countryside, where the appearance of the landscape depends on positive management carried out in a spirit of willingness and cooperation. Management agreements, rather than planning restraints, are therefore the order of the day (notwithstanding the desirability of having selective 'last resort' protection powers in the future).

The major difficulty in implementing a comprehensive plan for a rural area lies in the disparate and single-minded nature of the resource management agencies, whose objectives are geared towards national needs, and away from local ones, and which are administratively dislocated from the local government system. At the local level, the most favoured response to this situation has been to implement management plans for characteristic 'problem environments' which are associated with a high degree of conflict between users, such as the urban fringe, the upland and the coast. Commonly, project officers, often with little power other than that of their own talent to persuade, and who must be satisfied with the most limited of budgets, have been

employed as 'troubleshooters' by planning authorities. Such appoint-
ments are most effective in overcoming very local problems —
countering the effects of trespass and vandalism by the judicious siting
of new fences and gates, forging access agreements with landowners,
and coordinating the efforts of volunteer labour, for instance. The lead
for this sort of work has been given by the Countryside Commission,
who have made grants available from their experimental budget to
assist pioneer projects in such localities as the Lake District, the Bollin
Valley near Manchester, and the Glamorgan Heritage Coast (12).
Financial support to local authorities has been essential, as often
farmers will expect assistance for repair work to fences, stiles and so
forth, or the removal, into public management, of some of the
responsibilities of private ownership, notably the maintenance of
public rights of way.

Useful though such measures may be, however, they have largely
been *ad hoc*; whilst countryside management may often benefit from a
low-key approach, limited objective and the lack of a complete association with the local government system, it cannot prevail generally
without being couched within a more systematic planning framework.
To achieve this, the ground rules and operational context for management projects must be firmly established in structure and local plans,
perhaps requiring pressure to be brought upon the Secretary of State
to select conservation and farming as 'key issues' for the structure
plan's examination in public. Thus, Elson has argued that:

"There is an urban fringe problem, but land use manifestations of it
are not spread evenly around our cities. The situation remains
poorly documented, with too many small-area studies reflecting the
problems of analysis of what is a complex situation. Causes and
problems are less well documented than the end results of a myriad
of individual decisions. The urban fringe has suffered from being
seen just as a geographical phenomenon: its problems can only be
fully explained and solutions tested by study areas wider than this
narrowly conceived definition. Solutions will not be produced by
planners alone, nor by countryside management alone, especially
where it has no valid local land use planning framework in which to
operate" (13).

Ultimately, only central governments can initiate the rapprochement
of resource planning agencies necessary to ensure the proper management of the countryside. In response to calls for a Royal Commission
on the countryside, the government took the more cautious step of
establishing a Countryside Review Committee, which produced its
first report in 1976. This and its subsequent papers have made some
welcome recommendations for inter-Departmental liaison, but unfortunately these are not binding on anyone. Indeed Davidson and

Plate 18.   The urban fringe – a suitable case for treatment.

Wibberley dismiss its potential effectiveness on the grounds that "a Civil Service Committee, with its partiality for compromise and its unwillingness to rock the establishment boat cannot be expected to generate the kind of ideas about problems or their solution upon which discussion can build" (14). Perhaps the only way in which a change of attitude can be forced is by sustained pressure from planners; and the only way in which planners can achieve the necessary credibility to effect this is by their participation in management schemes, in which the confidence of the farming community is gradually secured.

Example 7: Ecology and site development – Llyn Brenig

Very often, it is appropriate to conserve or improve the vegetational cover of a site which is to be affected by development. Such action may range from small-scale landscaping around a single building, to the creation of completely new landscapes, such as at the Llyn Brenig reservoir in North Wales — at both extremes, a full consideration of the ecological characteristics of the site will ensure a greatly enhanced probability of success.

Llyn Brenig is an extensive balancing reservoir, some 4km long by 2km wide situated in the wild landscape of the Mynydd Hiraethog. From the start, it was felt necessary to consider the reservoir not just in its water catchment role, but in the context of its neighbouring uses of hill farming, forestry, conservation and recreation. Thus, the forward planning stage embraced both the broad issues of multiple usage and the detailed design of the conservation and establishment of vegetative cover. Ecological considerations formed an integral part of the overall development plan, for instance, requiring that moorland roads should be left unfenced and with rough edges where natural regeneration could take place, and that agreements with the Forestry Commission should be concluded to ensure that certain sections of mature forest were never clearfelled.

Varying degrees of success have been experienced in establishing vegetation on the site. Some rough grazings are patchy and others display rank growth, so that careful management will have to be applied in the future to improve the palatability of the sward; elsewhere, attempts to establish heather have been unsuccessful. However, in places the young forest plantations are displaying surprisingly high rates of growth, comparable to those usually associated with fertile lowland sites.

The creation of new landscapes, or the retention of old, on sites to be affected by development is, of course, a most testing exercise. On most large sites, a wide range of sub-contractors, often ignorant of landscaping techniques and careless in their use of machinery and storage of site materials, may be engaged. If satisfactory ecological conditions are to be retained, it will be necessary to attach the most carefully worded planning conditions and to ensure the most diligent site supervision.

Conclusion

These examples point to a broader role for local planning authorities in the management of resources. It is clear that environmental problems are too pressing, too inter-related and too technically demanding to be dealt with by a myriad of control authorities. The present approach to resource utilisation — narrowly departmental and often distinctly negative — which pervades both central and local government must be replaced by more positive, integrated management operated on an inter-agency basis. The efforts of those concerned with land use planning, highways, environmental health, forestry, conservation, water supply, health and safety, pollution control and energy development will be far more effective in overcoming our environmental crisis if they are united.

Perhaps more significantly, ecology may hold the key to the ideological revitalisation of a planning profession too often characterised by disillusion and expediency. In this sense, ecology must no longer be treated solely as a sub-discipline of biology but as the basis for a comprehensive environmental philosophy which, in one version, propounds that small is beautiful, that growth should no longer be the overriding principle of economic management, and that the winds, tides and sun provide better long-term energy prospects than plutonium. It is not a creed which finds universal acceptance, but it is one — perhaps only one of several which the ecological imagination is capable of offering — which provides a genuine alternative. It thus permits planners to present real choices to the community, and in so doing it not only enhances our environment but our sense of democracy also.

NOTES

(1) For a review of these issues see G. Smart (1978) 'Nature conservation and Planning', *Town Planning Review*, 49.
(2) Susan Teasel (1977) 'Planning and the Role of the Ecologist'. Unpublished DipTP thesis, Polytechnic of Central London.
(3) J. Elkington and J. Roberts (1977) 'Who needs ecologists?'; 'Is there an ecologist in the house?'; 'The ecology of tomorrow's world'. all in *New Scientist*, Nos 1075, 1076, 1078 respectively.
(4) J. E. Lowday and T. C. E. Wells (1977) *The Management of Grassland and Heathland in Country Parks*: a Report to the Countryside Commission prepared by the Institute of Terrestrial Ecology, Cheltenham, Countryside Commission.

(5) Marion Shoard (1979) 'Children in the Countryside', *Planner*, 65.
(6) C. Baines (1979) 'The conservation of urban derelict land', Conference of the Society of Chemical Industry, 1979.
(7) Country Borough of West Bromwich (1972) Structure Plan, Report of survey.
(8) A. MacLeary (1979) 'Energy development and land in the United Kingdom', *Planner*, 65.
(9) J. Longmore and J. Musgrove (1979) 'Energy conservation and urban planning in the United Kingdom', in Second International CIB Symposium on Energy Consumption in the Built Environment, Session 4, *Design for Low Energy Consumption*, Danish Building Research Institute, Copenhagen.
(10) M. Braiterman, J. Fabos, J. H. Foster and C. M. Greene (1979) 'Energy and land use: an assessment of the potential fuel conservation value of local landscape resources', *Journal of Environmental Management*, 8.
(11) A. Bell (1979) 'EIA and forward planning — the Cheshire experience', in J. M. Herington (Ed.) *The Role of Environmental Impact Assessment in the Planning Process*, Geography Department, Loughborough University.
(12) Refer to: A. Hall (1973) 'The Bollin Valley Project', *Recreation News Supplement* No 9; Countryside Commission (1976) *The Lake District Upland Management Experiment*, CCP 93.
(13) M. Elson (1979) 'Urban fringe management problems', *Planner*, 65.
(14) Joan Davidson and G. Wibberley (1977) *Planning and the Rural Environment*, Oxford, Pergamon.

Further Reading

Introduction

Some non-technical and thought-provoking reading about contemporary environmental issues is provided by:
R. Higgins (1978) *The Seventh Enemy: the human factor in the global crises.* Hodder and Stoughton, London.
Barbara Ward (1979) *Progress for a Small Planet.* Penguin, Harmondsworth.
A concise and lucid review of the arguments surrounding the ecological constraints on economic and population growth is contained in:
Open University, Earth's Physical Resources Course Team (1974) *Implications – Limits to Growth?* Open Univ. Press, Milton Keynes.
The philosophical and political aspects of the environmental movement are admirably covered in:
T. O'Riodan (1976) *Environmentalism.* Pion, London.
F. Sandbach (1980) *Environment, Ideology and Policy.* Blackwell.

Chapter 1

Further reading on the major physical systems of the earth is widely available, but some texts of particular relevance to the present study are:
L. F. Curtis, F. M. Courtney and S. T. Trudgill (1976) *Soils in the British Isles.* Longman.
J. E. Hobbs (1980) *Applied Climatology.* Dawson/Westview. (This text is particularly useful for its treatment of the southern hemisphere.)
Elizabeth Porter (1978) *Water Management in England and Wales.* Cambridge Univ. Press.
K. Smith (1975) *Principles of Applied Climatology.* McGraw Hill.
A. E. Trueman (revised by J. B. Whittow and J. R. Hardy) (1971) *Geology and Scenery in England and Wales.* Penguin.
J. B. Whittow (1974) *Geology and Scenery in Ireland.* Penguin.
J. B. Whittow (1977) *Geology and Scenery in Scotland.* Penguin.

Chapter 2

A very readable and non-technical discussion of nature is provided by:
R. Mabey (1980) *The Common Ground.* Hutchinson.

A straightforward presentation of ecological concepts is contained in:
D. F. Owen (1979) *What is Ecology?* OUP. 2nd Edition.
Somewhat more advanced treatments of the biotic system may be found in:
C. C. Park (1980) *Ecology and Environmental Management – a geographical perspective.* Dawson/Westview.
I. G. Simmons (1979) *Biogeography – natural and cultural.* Edward Arnold.
For specific discussions of agriculture and forestry, refer, for instance, to:
Sylvia Crowe (1978) *The Landscape of Forests and Woods.* Forestry Commission Booklet 44. HMSO.
Joan Davidson and R. Lloyd (Eds.) (1977) *Conservation and Agriculture.* Wiley.

Chapter 3

A variety of aspects of resource management are contained in:
J. M. Edington and M. Ann Edington (1977) *Ecology and Environmental Planning.* Chapman and Hall.
J. W. House (Ed.) (1977) *The U.K. Space: resources, environment and the future.* 2nd. edition, Weidenfeld and Nicholson.
D. Lovejoy (Ed.) (1979) *Land Use and Landscape Planning*, Leonard Hill. 2nd. edition.

Chapter 4

A brief, but very helpful review of pollution is:
K. Mellanby (1972) *The Biology of Pollution.* Edward Arnold.
Perhaps the most pertinent study of land pollution, which is of particular concern to planners, is:
R. P. Gemmell (1977) *Colonisation of Industrial Wasteland.* Edward Arnold.
Planners would also be well advised to keep up to date with the reports of the Royal Commission on Environmental Pollution (HMSO). Perhaps the most relevant of these are:
3rd. Report (1972) *Pollution in Some British Estuaries and Coastal Waters.*
4th. Report (1974) *Pollution Control: Progress and Problems.* (A revised version of this was published in 1979.)
6th. Report (1976) *Nuclear Power and the Environment* (The Flowers Report).
7th. Report (1979) *Agriculture and Pollution.*

Chapter 5

Some useful recent discussions of rural and natural resource planning techniques have been provided by:
J. T. Coppock and M. F. Thomas (Eds.) (1980) *Land Assessment in Scotland*, Aberdeen Univ. Press.
D. A. Davidson (1980) *Soils and Land Use Planning*, Longman.
N. Lee and C. Wood (1980) *Methods of Environmental Impact Assessment for Use in Project Appraisal and Physical Planning.* Univ. of Manchester Occasional Paper in Planning No. 7.

Chapter 6

For a general text on town planning practice in Britain, refer to:
J. B. Cullingworth (1979) *Town and Country Planning in Britain.* 7th. edition, London: Allen and Unwin.
More specialised texts include:
I. Bowler (1979) *Government and Agriculture – a spatial perspective.* Longman.
Joan Davidson and G. Wibberley (1977) *Planning and the Rural Environment.* Pergamon.
A. W. Gilg (1978) *Countryside Planning – the first three decades.* Methuen.
A. Walker (1979) *The Law of Industrial Pollution Control.* George Godwin.
C. Wood (1976) *Town Planning and Pollution Control.* Manchester Univ. Press.

Chapter 7

The best place to keep up to date with future developments is probably:
A. W. Gilg (Ed.) (1980 onwards) *Countryside Planning Yearbook.* Geobooks.

Index